CONTENTS

JUMPING IN

I'm on a journey to build the toilet of the future. The water flushing toilet that you are using right now is not going to last much longer because it has caused a problem of unprecedented proportions. We are facing a global shit disaster! Oh shit, really?! It's okay if you didn't know about it; I didn't know it either and I studied wastewater treatment plants in college. At first, I accepted the conventional wisdom that we should continue polluting the essential medium of life but then, one day while taking a shit into a composting toilet, it finally struck me: Shitting in drinking water is a bad idea; I had awakened.

The water wasting toilet is a modern invention. It came about in the 1840's. Prior to this innovation, people in the city used metal buckets. We've all heard about how the French used to throw shit buckets out the windows. *Guarde l'au*! Once I went to a toilet museum with a special someone and we saw the evolution of toilets from buckets, to porcelain bowls and eventually flushing toilets as water became available in households through metal pipes.

The toilet was always thought as a means to dispose "human waste," to carry it out of sight. No thought was given to treatment and the consensus at the time was that once the sewer was dumped into the local river, be it the Seine in Paris, or the Thames in London, then "dissolution was the solution." There was no scientific knowledge about viruses and water-borne pathogens. Wastewater treatment plants weren't invented until one century later. Washington DC wasted waters didn't get treatment until the 1940's.

Modern urban dwellers never saw any value in shit and piss. It's always been the opposite. I remember once after drinking too many beers with my friend Arnaud on the margins of the Seine, opposite to Notre Dame, and before we left, we peed somewhere near the water. Arnaud shouted, "*Vive la France*" before zippering up. When feces got in contact with water, cholera spread, and it was a major killer until scientists and health workers figured out it was the pathogens in shit that was killing people. However, by then it was too late, and we had already grown used to conveniently flushing our excrements away. Wastewater treatment plants have been playing catch up, but they have a long way to go.

You can add "going to the toilet" to the list of things that are bad for the environment. Do you feel guilty about eating meat and/or flying? I feel guilty about flushing the toilet, every day. According to the World Health Organization(WHO), two thousand people die daily from drinking water that was contaminated with shit, either their own or from someone upstream. Additionally, there are 4.5 billion people without access to sanitation which means their feces and urine are not being managed in a way that doesn't end up getting back in their water and food where it unsurprisingly ruins their day. Sometimes, I think that when I pull the lever and that waterfall runs down my toilet, a poor kid dies in Africa and it makes me feel horrible. In the last decade this number grew from 2.3 Billion. We're not making progress and I know the reason why. But what really keeps me up at night, is the fact that most of us think that the solution to the lack of toilets is to build more water flushing toilets. In my opinion, the water flushing toilet is the reason for our global shit crisis, but few know this and those of us who do aren't speaking out.

I used to think that we, as a civilization, had already gotten our shit together and it was only poor countries that needed to catch up. "Come on, third world peoples, clean up after yourselves," I whispered to myself. After all, its 2020 and Elon Musk is taking us to Mars next year. We are going to become a spacefaring civilization. But perhaps my own inability to take care of my own shit should have been an indicator that, collectively, we are not doing much better. I ignored these signs and preferred to "flush and forget" the fact that my own shit was contaminating perfectly potable water and polluting the local river, be it in New York City or Washington, DC. Looking back, I should have been more honest with myself. I should have listened to the wise words from my ex-girlfriend who told me *"You act as if your shit don't stink."*

It gets worse because we are running out of water. You know this. We've all heard that World War III will be about water, but what you may not know is that it has started already. Again, there are *water scarce* and *water stressed* cities and countries quarreling with their neighbors over water. Egypt and Ethiopia, for example, are fighting over the Blue Nile river waters and a proposed damn to secure irrigation water in Ethiopia. Guess what Egypt thinks about limiting their water supply? Chennai in India, Cape Town in South Africa, California in the United States have registered alarmingly low water levels. Cape Town came days close to what they called "Day Zero" which would be when water would run out. If you live in the outskirts of major cities like Chennai or México City, you are already out of water. Tragironically, when there are forests in place, then rain falls from the sky—naturally and free of charge. But we've messed the entire ecosystem and that's why I feel guilty when I flush the toilet, twice so all the shit really goes away.

There is hope, however, and it's inspiring enough that it moved me to great lengths, 15,200 meters to be exact. I've met a few other enlightened individuals who have been leading the way and have showed me that there are alternatives to the water flushing and evil porcelain toilet. These heroes have written books so they know a lot about shit, and they've given me advice, sold me their urine-diverting seating pans and are now helping me build my own toilet of the future that will one day solve all of our shit problems with one simple button. I can't wait until my startup can sell it on *Amazon Prime* so everyone can save the world with one-click.

This toilet of the future is so amazing that it will save us from our own shit **WHILE** turning shit, something traditionally considered bad, into humus, something everyone knows is good. It's a win, win...win for us, the environment and the shit. This may be a hard one to swallow but there is value in shit. I know, I also couldn't believe it, but I read in those books and I saw it with my own eyes (smelled with my own nose, in fact) that excrements, the stuff that comes out of our bodies, when turned to compost, have amazing nutrients. I mean, sure, shit stinks, after all it comes from our assholes, but it still is good for something if composted properly. I also know that feces have bacteria and pathogens that kill people but using hundreds of liters of drinking water, the universal solvent, to transport our shit from our homes so someone else can deal with it doesn't sound like the most reasonable solution to me. Here's the way I understood it: In nature, there's no such thing as waste. Only we, Homo Sapiens, the intelligent ones, create waste. If an elephant shits in the forest, then its dung is not trash, it's food for dung beetles. How about oil? Is it a waste or a resource? It depends where you put it. An oil spilled in the ocean is obviously polluting but oil piped into a refinery is considered a valuable resource. How about carbon from internal combustion engines running on fossil fuels? Again, it depends where the resource is. Carbon in the atmosphere pollutes whereas carbon sequestered in the soil sounds like music to environmentalists' ears. This toilet of the future knows and understands that excreta is a resource. It's incidental if it sounds like this toilet is an intelligent being but it's truly a special technology. It's not artificial intelligence, it's natural. As you can tell, I'm thrilled about this toilet and I can't wait to share what I have learnt from reading those books, working with shit heroes and cleaning a lot of toilets at England's best music festivals.

The best thing about this toilet of the future, as I mentioned in my TEDx is that it already exists. In fact, so called "savage tribes" from México had mastered this technology but we've lost it through wars and conquests fueled by arrogance and its counterpart, ignorance. We used to be able to take care of our shit by ourselves, but something went wrong in history and all the shit got loose. Thankfully, however, for our shared benefit, these ancient peoples, long-gone and almost forgotten, left behind clues which I have been puzzling together. I couldn't find much in the history books because, apparently, victors don't like to talk about the good shit their opponents left behind before they were killed. But the evidence that I have found is encouraging. The rest, we will have to surmise, conjecture and test for ourselves. The good news is that I feel as if the shit gods of ancient times have chosen me as the one who will rescue their glory and reveal this long-kept secret to you.

But wait, there's just one more problem. I don't have the toilet, yet. I mean I built it and I gave it to people who I thought needed them but in the end I'm not sure if they wanted it, so I want to build a better one. That's where we are right now. It's taking me a while because my own shit is getting in the way of me helping the world take care of their own shit. It's very complex. My laziness and procrastination, experiences and travels around the world are slowing me down.

However, there is more hope. I'm one of the most persistent persons out there. Look at the title of this book, I swam from Spain to Morocco to build prototypes of the toilet of the future in México. That's gotta be worth something. It was a 15.2 km (9.6 miles for you norteamericanos out there who need to get on with the metric system because I'm tired of converting units) athletic endeavor where I proved to myself that I have what it takes to build the toilet in which you will want to crap next. And I did this (I swam between two continents even though few were watching, hence the need to write this book and tell more people about it) because I feel this is the toilet that is going to save us from ourselves, as Karl Sagan eloquently said it. I'm also not afraid of asking questions even, seemingly stupid ones, as you will read here. I've accepted so many things as truth, but the shit showed me it was time to question everything in order to solve this problem.

Sorry, but there is actually one last *one more problem* ...It is you. You are in the way of the toilet of the future. And I don't mean that you're in line to use the toilet of the future, I wish that was the case. I mean, you're also the reason the toilet of the future hasn't fully arrived yet. But it's O.K. because I was once there where you are. I was ignorant, like you. We won't be able to solve the global shit problem until you take a good and honest look at your shit as well. That's why I couldn't not write this book. Trust me, writing a book is hard, but not writing a book about something which you are passionate is even harder. It was impossible not to put these words together. It had been haunting me since I swam across the Strait of Gibraltar on November 3rd, 2017, I will never forget the date. I was moved to write this book so it becomes our manifesto and you can join us in the *Revolootion*, to quote one of the shit heroes you will meet later in this odyssey. That's why I call it the global shit problem. Everyone shits; therefore, everyone needs to get involved.

Shit is the one thing that unites all of humanity. No matter where you were born, your gender, ethnicity, nationality or skin color, we all shit. Some of us squat, while others sit but we all need to evacuate our bowls. Sure, shit may have different consistencies, colors and fiber content but we all need to do it, every day, no matter if you're a homeless guy in the streets of Denver or the Queen Elizabeth in the Buckingham Palace. We can tray and separate ourselves as vegetarians, vegans or pescatarians but no matter what living statement you're making about the food you put in your body, all of us equally defecate and urinate. What we do with our shit, will define us for posterity. We will either be known as a shit-fearing civilization or the generation who got their shit together and solved the global shit problem. In which side of history do you want to be? I'm assuming, not-the-shitty-one.

Let's get this shit straight before we start. We're in the middle of a global fecal problem that you didn't know existed and if you were asked to solve it with the conventional solution then, you'd actually be making it worse. The reason for this problem is the water flushing toilet because it has silently killed more people than wars, wasted two valuable resource and alienated us from our own human nature. The solution to this shitty nightmare of unprecedented scale is the toilet of the future which already exists, and it's being revealed to us by an anciently wise tribe. The challenge will be to build a working prototype of this toilet to convince you and the rest of the world that we have to change the way we shit, or shit will kill all of us. And to build this toilet, I will do everything, even swim across the Strait of Gibraltar.

Wake up because we are going for a swim! Do you like jumping in headfirst or divebombing?

1 – BUBBLES

Eu sou o filho d'Água

Once my mother told me I was conceived next to a waterfall in the *Cerrado*, the Brazilian savanna. Brasília, the capital of Brazil can be a pretty boring place, but it's surrounded by beautiful landscapes, and sexy *cachoeiras*. This intimate revelation explained my love of swimming. Appropriately, one of my earliest memories is of swimming in my local sports club called SESI located in my hometown of Sobradinho-DF. I must have been five years old and my swimming category was called *Piabinhas* which is the diminutive of minnows, as if it needed one. The instructor would tell us to step inside the ankle-deep pool, lay on the bottom, belly touching the floor and hang on to the edge so that we could lower and raise our heads in and out of the water. She would then instruct us to make "bolhinhas," or small bubbles, for the entire class. That was the one and only thing I learned as a *Piabinha*, making bubbles. It didn't' take me many lessons to learn how to push air from my nostrils underwater. I think I must have mastered it quickly. I would obediently lower my head and watch the bubbles float to the surface. I also remember blowing out so strongly that my head vibrated as the air bubbles exploded underwater. The exercise continued, going up for air, inflating my lungs, and lowering my head in the water while pushing the air out of my nose; inhale, exhale, and repeat ad nauseum. I also remember looking over, every time I came up for air, to the deep pool where the older kids, the "Dolphins" were swimming and I hoped that one day I would swim with them.

I've been swimming ever since; it has kept me physically and emotionally healthy. Whenever, I spend too much time out of the water, I notice I tend to become anxious or sad. When I first jump in the water, and start swimming, I feel as if I were a newly built ship on its maiden voyage. The swimming cap keeps my head warm. The earplugs drown out the outside noise and for a refreshing moment, I can hear my own breath again. Ritualistically, I lick the inside of my goggles to temporarily prevent them from fogging. The freshwater shocks my skin but after I get in the zone, also known as flow state, I can't tell the difference between me and the liquid flowing around. In the water, almost completely free from the restraints of gravity, my body floats and my thoughts are liberated. Some of my *best ideas* have come to me in the water. Swimming is the closest I can feel to flying.

There is a bird called the Gannett that knows both of these feelings. From heights of 30 meters it dives at speeds of 100kmph stretching its neck and folding back its wings so that it penetrates the water surface like a missile. If the bird opened its wings, thus increasing its surface area, then it would probably kill itself because hitting the water at that speed with such momentum would be almost like hitting the ground. But the Gannett knows the physics behind water surface tension and makes its body thinner like an arrow so that it can pierce the thin ocean layer which would otherwise destroy it. Once its dove past the surface, reaching depths beyond most divers, and the kinetic energy from its dive has been transferred to the water particles, then the Gannett is once again free to open its wings and "fly" deep in the ocean.

After learning how to make bubbles, I learned to streamline and proper swimming technique. Streamlining means to stretch your arms as much as you can, elongating my limbs like the Gannet. Personally, swimming technique means to have awareness of where each body part is at any given moment during the stroke cycle and making sure they are where they need to be. It's not easy, at first, to know where to place your right hand in the water while simultaneously thinking about hip rotation and kicking your legs. It took me years to become aware of my bad technique and I assume that it will take me a lifetime to perfect it.

Swimming during college also kept me in healthy friendships, away from partying. Even though I was going to college in a party town, I didn't really go out. I didn't join a fraternity in school. I was always studying, working or swimming. There was a time when I had swim practice at five in the morning. Weekends were also consumed by longer training sessions and the accompanying "swim meets" which are competitions. When I did go out on a Friday, I insisted on going home before midnight which earned me the nickname of Cinderella. Swimming also helped me structure my life schedule. Practices would be either early in the day or evening, university would take place in the middle and work afterwards. When my schedule changed, I always managed to find time to swim, even while traveling. I've swam in European and Mexican pools. Once, minutes after landing in Manchester, England, I found myself swimming in the city's Olympic swimming pool. Swimming also provided me with a close network of friends such as late dear friend Walter Kramer, a German doctor who hosted me in his house when I was traveling through Germany looking to learn more about a German engineering company.

In the Florida swimming circles, I had frequently heard about the swim around Key West. It's a 20km (12.5 miles) swimming race around the island which is the last stop in the Florida Keys. In order to get there, one must drive down four hours through a chain of islands connected by bridges and narrow land strips.

Key West, the southernmost point in the continental U.S., is only 90 miles (144KM) miles away from Cuba, they like to mention. The island is a tourist destination famous for its beaches and Caribbean-like weather most of the year. The race had always captivated my imagination and one day, I convinced two friends Félix from Nicaragua and Márcio from Brazil to support me on the swim. Long distance swims like this take more than four hours so I need to stop and eat; it's called feeding and they would row in a double kayak with our food supplies. Their job was also to make sure I was safe and swimming in the right direction which are just as important as eating. I rented a car, booked a hotel and drove us down the 160-mile (260 km) drive from Miami. We left on a Friday after my work.

At the time, I was doing a long internship at the Fort Lauderdale/Hollywood Airport Runway expansion construction project. This internship had been a dream come true. I first discovered Odebrecht when they built the FIU Football stadium. The job was on my way to my pool at the Miami Aquatic Center (MAC). I saw the company name on a banner in bold red letters and googled it only to be surprised by the fact that they were a Brazilian engineering company working on my campus. At the time, I had just started studying civil engineering in addition to finishing a degree in Public Administration.

I had always been interested in government. My uncle and aunt both ran for public office and were elected in the interior of Brazil. They dedicated themselves to improve the lives of less fortunate Brazilians. After I moved to the United States when I was fifteen, I started thinking about what I wanted to become when I grew up. The cultural shock opened my eyes to the disparities between rich and poor, developed and undeveloped. The experience strengthened a notion that I held that one day I'd come to Brazil and do something to help others. That's why I was studying Public Administration; I wanted to come back to Brazil after my studies and join civic life.

However, I was almost done with P.A., but I felt as if I was cheating myself. I understood government but it was not enough. I felt as if I was wasting the opportunity of studying abroad and I wasn't pushing myself to the best of my abilities. I was getting good grades without studying. This political stuff was too easy. I had just spent a semester interning in the U.S. Congress, General Motors and studying at George Washington University in Washington DC. I also volunteered for Ron Paul's campaign, a libertarian candidate who I admired for amongst other reasons, his staunch stance against wars. Ron Paul didn't go very far, and I remember my time in congress mostly by the happy hours set up by different lobbying organizations advancing their own interests over the public welfare. I became disillusioned with politics and returned to Florida.

It was then that I decided to study civil engineering, in addition to finishing P.A. I always read magazines like "The Economist" and books on economics and development out of curiosity. Some books like "The Dictator Handbook," and "Why Nations Fail" stood out to me. I was beginning to see the need for not only political reform through a transparent government, but I also felt that countries needed infrastructure to grow. I looked at the well-maintained highway system in the United States and compared it with the Brazilian counterparts. I could only conclude that for Brazil to be a first world country, it also needed highways, ports and airports.

Back then, Brazil's economy was growing, and its president Lula was the most popular politician in the world. The Economist magazine published an issue with the statue of Christ the Redeemer, Rio de Janeiro's symbolic statue, taking off with a smoke trail like a NASA rocket. Even U.S. President Barack Obama rendered him compliments, calling him "The Man." It seemed that Lula and Odebrecht were riding the economic wave and lifting millions from poverty by brining construction projects to Brazil. Odebrecht was going to be my ticket back to Brazil. I felt like I was at the right place at the right time.

We arrived in Key West and ate pasta at a local diner to load on carbohydrates I would need the next day. Carbohydrates are how we get energy from our food; they're nature's candy bars. In the race morning, we had overcast skies with strong winds blowing the palm trees backwards. After having lived in Florida, I grew used to flash tropical storms which and hoped this was another one and it would soon pass. I went to the swimmer sign up area, got an orange flag to strap on to the kayak for visibility and a volunteer wrote my race number with sharpie on my arms and legs. At the beach, all the other kayakers were doing last minute preparations and securing their coolers while the swimmers were rubbing up a zinc paste to their bodies, making everyone look ghastly white. Conventional sunscreen will not last in saltwater for long, and since the swim was expected to last at least five hours for most of us, then zinc one can buy at the pharmacy was the best sunscreen. Zinc is white, reflecting the sun, and thick sticking to the skin for days. I asked Márcio to apply some to my face, back and shoulders, impregnating his hands with the white paste, while being careful not to touch my googles, blocking my vision.

It was almost time to begin the race and the other swimmers started walking over the seaweed blanketing the shore. The kayakers dragged their boats into the water and jumped in. My kayakers were vital to my swim. They would keep me safe, help me navigate and stop for refueling so I could swim the 20km. Communication between us would be critical. I tried to go over with the details with them, but it was becoming clear that Márcio's Spanish would not be good enough as was neither his English. Luckily, Félix was very patient with him, speaking Spanish and English but no Portuguese. *Derecha* and *Direita* are almost the same, I told myself, hoping for the best.

I made the Christian cross sign that I learned to make during scary situations and walked across the algae mat on the beach into the water. The waters of south Florida are warmed by the current coming from the Gulf of México bringing the temperature high enough as to not need a wetsuit; I've never felt it drop below 15 c (50F). I swam in and looked back to Félix and Márcio getting the kayak in the water. They seemed to struggle, taking a long time to get in and row towards me. I waved my hands in the air to show them where I was, but they had better things to do like learning how to row in a straight line or synchronize who paddles on which side. I couldn't hear from the distance which language they were speaking or if they understood each other at all. As I waited for them to get organized, I was forced to "egg beat" which is something I learned while taking water polo lessons at Boca Raton when I was in High School. Eggbeating means to kick your legs in a circular position like an eggbeater, allowing a polo player to stay upright and upon more kicking power, raise above the water line to receive or throw the ball. The first half of polo training sessions was only eggbeating, similar to making bubbles for an early swimmer. I was pretty bad at water polo and didn't continue training but at least, learning how to egg beat had paid off.

I met Félix in the Miami bike scene. I'm not sure if it was during Critical Mass rides or Bike Polo sessions because we were active in both. Critical mass rides take place in hundreds of major cities across the world. You can join them by bicycle, roller skates or any human powered vehicle. They normally take place once a month and tend to be self-organized. People get together online and decide on a meeting place; the routes can be either pre-planned or improvised. The objective is to ride through the city *en masse* which unavoidably causes traffic disruptions. It shows car drivers that cyclists and other means of transportation also have the right to be on the streets. I've noticed that I enjoy any city where I can cycle.

Bike polo is similar to horse polo but with bikes. I learned to play it when I lived in downtown Miami and Brickell. We would gather afternoon on Sundays beneath the I-95 overpass along the SW 8th street or Calle Ocho, on the edge of Little Havana, a neighborhood in Miami. We played it mostly with fixie bikes for better control although any bike would do it. We used mallets to hit the hard ball in between the goal posts. I loved those days in the Miami bike scene. Many of the cyclists wore cycling caps or "casquettes," famous among racers.

Márcio and I had been neighbors in Deerfield Beach, a city in South Florida with a large Brazilian immigrant community. He is from the interior of Brazil and was doing construction jobs like most undocumented male immigrants to South Florida. He was an experienced tile layer, making at least four times what a similar position would pay him in Brazil. After long and exhaustive days, he would be found resting on his couch watching Brazilian Football League games of his beloved Flamengo.

Meanwhile, the boats, swimmers and kayakers were swarming around me. Marine fuel exhaust in the air was making me nauseous so I swam away from the boats with their propellers which can easily cut swimmers' limbs with numbers written on them. Finally, my kayak dream team arrived, and I pressed how important it was for them to stay near me so we wouldn't get lost. If that happened, then I would have to get out of the race. They replied in agreement in their respective languages and after some minutes waiting, a horn blew off in the distant beach and we started. It took me a while to find my pace and for the two in the kayak to learn how to row together. I suspected it must have been Márcio's first time rowing. They kept rowing far away from me, either too slow or too far and it was driving me crazy. They were supposed to stay near me. What if a shark ate me? The strong winds were creating waves and the water was turbid resulting in no visibility under water. I was used to swimming in clear waters and the inability so see anything beyond my fingertips was making me anxious. I'd yell at them to stay near, but I don't think they were communicating effectively nor was I sure if they could even hear me.

They were supposed to help me navigate, choosing the fastest way around the island but instead they were going way off course, beyond most swimmers. They were in essence, making me swim longer than I needed. I tried to breath, making bubbles to calm myself. "Come back here!" I'd shout while eggbeating. Then, it came time for the first feeding, and I grabbed an electrolytes bottle and fresh orange segments from the cooler. I wanted natural sugars to fuel me in the swim and had read it would be a good flavor against the saltwater.

But three hours later, it wasn't. The citrus was exacerbating the salt in my tongue. Four hours later and nothing that I ate or drank could wash away the salt from my tongue which by now was swelling. The swim was taking me longer than I expected. I didn't know where I was, and it was hard to know how many kilometers I had swam. Periodically, Félix would get updates from other kayakers who passed us because I was so slow, but the numbers weren't accurate. I was growing tired and started feeling pain in my right shoulder. Why did I want to swim 20 kilometers, I kept asking myself. I must have thought of this while doing a short swim. The weather didn't improve but at least they weren't going to cancel the event although I must have wished for that several times. I even thought about quitting but the thing about swimming around and island and then quitting halfway is that I would have to walk the rest of the way around the island. How would we carry the kayak? It looked heavy.

Still, the persistent part of me told myself to enjoy the swim, forget the pain and make bubbles. We stopped for fueling at hour number six and Félix kept telling me that we were almost done but I couldn't tell where we were, and I had been hearing that for the last hours. To make matters worse, I didn't want to eat anything I had brought to the kayak. By now my tongue was swollen and burnt from the salt but then I remembered I had brought a chocolate brownie flavored protein bar from Clifbar. This protein bar was my favorite. I had tried Powerbars before, but they tasted like pure chemicals. Clifbar's were the only one I could eat more than once during the same training and the chocolate actually tasted great with sea salt. I ate it and kept swimming.

This Clifbar, a lot of persistence and my Kayakers were what I needed to finish the swim around Key West in 7:15 hours. When I stumbled on the beach and got a tiny medal, I didn't even care that I had finished the race second before last, after many swimmers 50 years or older. It was one of my proudest moments up until then. Little did I know how important that lesson of not giving up would become.

Outside of school, when I wasn't swimming, I was reading books about swimming, triathlons and all kinds of subjects. One of the coolest books I've read is called "The Lives of Ants" by Laurent Keller and Elisabeth Gordon. I literally could not let this book down. I didn't know ants could be so interesting and vital to their ecosystems. I also read biographies by athletes such as Chrissie Wellington, a British female Ironman champion who taught me a lot about determination in sports. Once I picked up a book by long distance swimmer Lynne Cox who used her swims for political goals like easing tensions between the United States and the Soviet Union during the Cold War. Her efforts opened my eyes to using sports for change. She also talked about swimming across the Strait of Gibraltar and finishing the "Seven Channels," a global circuit of seven channel swims, one of which is Gibraltar. I was so inspired by her accomplishments that I decided I'd also swim for a political change. I just didn't know what at the time.

Once I finished the book, I googled the seven channels. There was the famous English Channel and others such as the Malakai in Japan, but they all seemed too long, dangerous or too cold for this Miami swimmer, so I settled on the Gibraltar Strait because it was the shortest, "only 15km" and the warmest at 19c (66F). I researched and read more from swimmers who had successfully made the crossing and all of them had contacted the local organization called *Asociación Cruce a Nado Estrecho de Gibraltar*, which, at the time, was run by Rafael Gutierrez Mesa. To be considered for the swim, they asked for proof of having swum a similar distance in open waters. I first contacted Rafael on the

20th of May 2015 to inquire if I could swim the Strait that summer, informing him that I had been competing in South Florida triathlons, and that I had completed the 20km swim around Key West. I thought I met the requirements, but he replied that all the slots for the 2015 season were taken. However, he could put me on the waiting list for 2016 with a caveat that it would still be hard to get a spot because there were many requests from previous years. Swimmer placements would be done in November, he told me. Then, I started training even harder, and each swimming session in the pool, or on Sundays in Miami Beach with Venezuelan friends Evanova and Diego became focused on the goal of crossing Gibraltar by the end of the summer of 2016.

A lot of people ask me how I got into toilets. They want to know what moved me to do this dirty work. For the longest time I didn't know what to say. I'd mention I studied engineering and I liked swimming, but no one really got it. I should start saying that I like to swim in water without shit in it. But once I told a personal story that goes like this. In Brazil, my parents made a picture album/baby journal where along with the pictures, they registered special events in writing such as when my first tooth came out and my first spoken words. Years ago, while browsing through this album, I noticed one special entry in tiny handwritten letters, as if my aunt who wrote it was too ashamed to record that when I was 9 months old, "baby Aldo ate his own poop."

2 –SURFING

*"You have friends all around the
world, you just haven't met them yet"*

Couchsurfing motto

Back in 2014, I re-joined Couchsurfing which is a social network that connects travelers and locals. I had created a profile in 2012 but I never got around to using it. This time, my former Peruvian flat mate, Efra showed me how it worked. He was hosting travelers and practicing some of the four foreign languages he speaks. I had always been impressed by how he learned to speak Portuguese by playing online video games with Brazilians. He spoke it fluently despite never having set foot in the country. Now he was hosting French travelers and speaking that language too.

I first started learning English as a child in Brazil. I had an uncle and aunt living in the United States, but it wasn't until my parents divorced and my mother went to live and work there that I decided to learn English. When I was fifteen, I moved to Boca Raton, an affluent city in South Florida. Although I had a good grasp of English grammar, I wasn't comfortable with speaking it because no one understood me. Every time that I tried saying something, the other person would never understand me, and I dreaded having to repeat myself and not be understood again. I don't know if it was my pronunciation or if I wasn't loud enough, but I was anxious about each interaction with Americans. The first months were challenging but luckily, I was in the company of other foreigner kids. When I arrived in high school, they placed me in English as a Second Language or ESOL classes along with other students from Latin America and indeed, the world. I made a friend from Russia called Yuri and met other Brazilians and fellow Latinos. There were so many foreigners that I ended up learning Spanish before English. I just had more opportunity to speak Spanish with all the Colombians, Mexicans and Cubans living in South Florida.

I eventually became fluent in both English and Spanish and these experiences gave me the confidence to continue learning more languages. I was going to use couchsurfing to learn more. Couchsurfing works by allowing local "hosts" to welcome travelers or "Couchsurfers" into their homes. You don't to have to have a spare room, the couch is fine, even the floor is acceptable. Budget travelers are not a picky bunch. Miami hotel prices were so expensive that most surfers were happy to sleep on the carpet or their sleeping pad. At first it seemed like a dangerous proposition to host complete strangers but after you meet or host someone you can leave a personal reference and this feedback allows us to get a glimpse of the Couchsurfer or host whose couch we are surfing. When I mentioned the site to friends, the first question they asked was: "Aren't you afraid they will steal your stuff or kill you?" I replied by saying the surfers could suspect the same from me. The beauty of couchsurfing, and traveling at its core, is to trust people and the road. I learned to read surfers by their diction, photos and messages. I learned to trust strangers. I saw that the vast majority of people in the world are mostly good, yet we are all afraid of one another. What's the worst that could happen?

I filled out my profile, uploaded some pictures to let my prospective surfers know a bit about me and where they would be sleeping. Efra wrote my first reference and with his blessing, the requests came immediately. Miami is a tourist destination but at the time, there was only one expensive hostel in South Beach. I learned that European travelers were used to staying at inexpensive hostels, but they couldn't find any in South Florida. In fact, I researched hosteling and learned that Germans invented the concept as we know it, low-cost and simple shared accommodations for backpackers. Additionally, most Europeans on their way to south America wanted to spend a couple of days in Miami, on an extended layover. I understood why Miami was called the gateway to the Americas.

Meeting European and international backpackers shattered misconceptions I had ignorantly held for years. First, I realized not every traveler was rich. I had erroneously assumed it took a lot of money to travel but with each traveler I hosted, I learned that one could also travel on a budget. One Danish guy I hosted sported a beat-up, pink, flip-phone and second-hand, worn-out clothes. He was not what I imagined the Danish would be like. He gave me some of his Snus which is a powerful smokeless tobacco that is common in Denmark. I met real people on inspiring personal journeys around the world and I wanted to help them reach their goals. Offering my empty couch and letting them use my shower was all I needed to do to help.

Another benefit of hosting these surfers was being exposed to great new music. I confess, I had a bland music taste before traveling and meeting my surfers but now I love my playlists. I started listening to electronic music, psychedelic trance and a lot more. It's amazing how much more music there is out there, beyond what you can access on the local radio or TV. Music from other countries is like food, and culture in the sense that it is beautiful because it's different from ours. Since I've experienced so many flavors and sounds, I've come to appreciate diversity, difference and chaos.

It was awesome to receive requests from random people from places of which I had never heard and within hours have them in my living room, learn about their journeys and be inspired by them. Once I hosted a traveler, the first at the time, to do a trip around the world solely with Bitcoin. This was in 2014, a long time before it was well known. He gave me five dollars in bitcoin which would be worth one hundred dollars in 2020 but I lost my bitcoin wallet when my phone was stolen in the Mexico metro. I also hosted a couple from Netherlands who were working together. He was a journalist and she was a photographer. They were able to take their relationship on the road and it was beautiful to see their commitment to one another.

I was impressed with how quickly I became good friends with my surfers, sometimes as soon as they entered my house. Some brought gifts and cooked meals for us to eat while they shared their travel stories from around the world. I met solo travelers, couples and recent grads on their gap year, taking some time off after college to figure out what to do with their lives. I hosted hippies and fancy Frenchies. So many people went through my house that my good friend Mike's father asked me if I was running a human smuggling operation from my apartment. I hosted two Lithuanian girls backpacking and one of them, Rasa, told me about Burning Man, an art and music festival that takes place in the Nevada desert. She couldn't really explain it, but it caught my imagination and I decided one day I'd go there. She also told me about TEDx talks and how she volunteered in one. Hosting these incredible human beings in my living room was opening me up to the world and I couldn't have enough of it.

I was getting ready to eventually travel full time, so I wanted to earn credentials and experience by hosting travelers in my apartment in Miami. The strategy worked better than I anticipated. I met and hosted hundreds of travelers from around the world for two years. They challenged my world view and shared theirs with me. I learned that one didn't need a lot of money to travel, my "surfers" taught me a lot of their resourcefulness. I found out there was a whole network of passionate globetrotters willing to share their couch and a bit of their day with one another. But the most important lesson I learned from hosting travelers in my apartment in Miami, was to trust strangers.

Meanwhile, I continued participating in local short distance triathlon races in Miami. In case you didn't now, triathlon is a sport which combines swimming, cycling and running made famous by the Ironman triathlon event where athletes swim 3.86km (2.4 Mile), cycle 180 km (112 mile) and run 42.2 km (26.2 mile). This sport is ridiculous, and it attracts the type of personality like mine. People who want to prove themselves, to themselves and why not, the world. Additionally, it's an expensive sport, race entries cost around $150USD and the training and gear, including multiple thousand-dollar carbon fiber bicycles pile up quickly. Sometimes, I volunteered with Multirace, the local triathlon races organizer in Miami, in order to get discounts on the fees.

I did a humbler distance called Sprint which is 750m (0.5mile) swim, followed by a 20km (12.4 mile) bike ride and a 5km (3.1 mile) run. I first gave it a try in 2008 after being inspired or challenged by sibling rivalry since my brother competed in many races. Back then, triathlon was an obscure sport, but I think it suited my personality, more than my student pocket. As opposed to team sports, all the glory and responsibility would be mine. When I was growing up in Brazil, I used to play goalkeeper and I didn't miss the burden of losing a match and the reproaching eyes of my teammates whenever the ball hit my net. Triathlon was different, I was responsible for my own performance and I could train whenever it suited my schedule. Being a good swimmer gave me the confidence to enjoy the beginning of the race unlike most triathletes who come from running and are terrified of the water start.

I had spent some time out of the water and when I went back, I started getting panic attacks during a simple 25 meters lap. I'd start swimming by pushing off the wall, then start breathing on my right side, my initial breathing side, then start moving my arms. First, I would push off water with my right arm close by, bring it up and breath to the right. All the while, my left arm would be getting ready to enter the water and start the arm stroke cycle again. But before I could do at least three cycles, I'd look down my lane and see how far I still had to go, and I would panic. I felt like I was running out of air. I wasn't breathing enough because I was more concerned about how far I still had to go then how I would actually get there.

Counter intuitively, in order to get a full lung of air, I must first exhale all the air from my lungs, that is, making a lot of bubbles. Then, with an empty lung, I could fill it back up with fresh air every time I came up to the surface. It sounds weird but the water taught me how to breath. Perhaps, it is similar to our first breath of fresh air once out of the womb. As fetuses, we were just swimming in the placenta liquid, getting dissolved oxygen from our mothers but then, suddenly, we must breath this other fluid called air.

I kept trying but I'd have to stop and stand up in the middle of the shallow pool, incredulous of my inability to finish a mere 25 meters. I remembered once I invited my college buddy, Rei to come to the pool with me and while he was chilling, I was nervous, but his presence silently encouraged me to try again. I feared I'd never be able to swim again. And then, I remembered that I had learned how to swim a long time ago in the womb and I tried again. Heck, I had won my first race in life as a sperm. I thought of the bubbles lesson as a child and I told myself to breathe every three arm stroke cycles. I think the three cycles worked for me because it was an odd number which allowed me to alternate sides, thus not favoring any one set of muscles. Also, inhaling less frequently, allowed me to focus brain power on making more bubbles and on my stroke technique, hand placement in the water and kicks.

When you're exercising, your brain is consuming a lot of energy. Even at resting heartbeat, the brain is the most power-hungry organ in our bodies. When we are exercising the brain is consuming even more energy in the form of sugars. That's why that Clifbar allowed me to swim around Key West. In sports, when we are running out of energy, athletes call that "Hitting the Wall." I've felt it once before. I went out on a long bike ride to Homestead on a Sunday morning, without having slept well nor getting enough food for the road. The total ride must have been about five hours and I had to step off the bike during the last 500 meters and walk the rest of the way. I was so weak that I was about to faint, and I saw one lady watering her plants and I asked her for food saying my blood sugar was low. She who turned out to be diabetic, ran inside her house and brought me a chocolate bar and ice-tea. It was incredible how I felt as soon as I started eating the bar. It was if I had fueled up my fuel tank and the gauge needle quickly moved up. I started eating better after I realized that food is fuel. My diet became rich in carbohydrates from pasta, potatoes, rice and beans. Bananas, apples and oranges were my daily essential fruits and I ate oatmeal with strawberries and honey for breakfast.

I got my rhythm back, started swimming regularly and joined the MAC Masters Swim Team coached by Kirk. However, just breathing was not enough. One day Coach Kirk made fun of my rotation to breathe. He asked me if I was breathing from my naval because I was over-rotating. Swimmers need to rotate their hips so that we can extend the arm that is going into the water. Meanwhile, we must turn our heads to the opposite direction and the wake created by our heads will clear the water from in front of our mouths so that we can grasp a lung full of air. Then as the hand that entered the water goes down and pulls, we exhale air from our nostrils, making bubbles. All of this needs to happen while we are kicking our legs. I never forgot Kirks comment and became more conscious of my rotation, saving energy.

2015 was turning out to be one of my most active triathlon seasons. I joined a team Try2-1 coached by compatriot coach and runner Marcelo and saw a substantial improvement to my running technique, up until then my weakest discipline. It was also motivational to train alongside other athletes and enjoy some sportive comradery which I missed. One of the aspects of triathlon that I love is that it does you no good to be a fast swimmer first out of the water and then run out of fuel during the run because the slow swimmers will catch up and pass you. Instead, it pays off to be consistent across the three disciplines. I had a knee injury in 2012 from overtraining for a half-marathon which had made me anxious about my running. I would end up losing to the same athletes I had passed during the swim. It was frustrating and I looked to Marcelo for help me with my weakness. I remember going to the running track at 5:00am and training with the other triathletes. It was so early and dark that once we witnessed the smoke trail of a SpaceX Launch, although at the time it looked like an UFO. SpaceX is a rocket company started by Elon Musk.

At university, I struggled with one class in particular and, in hindsight, I should have paid more attention to this clue. The course was called Introduction to Environmental Engineering, ENV3001, and it was taught by Dr. Shonali Laha who was from Calcutta, India. I still remember her sweet-sounding voice and peculiar "Indian head wobble." When Indians speak, they tend to wobble their heads making a figure "8". At first, it's confusing because it's not the usual nod from left to right or up and down. It seems their necks have more degrees of freedom than mine. I remember being both amazed and confused because this head wobble can send mixed messages like "maybe" while they are verbally saying "No."

Dr. Laha's class had three main questions, one of which focused on Wastewater treatment plants or WWTP's. She introduced us to the concepts of primary, secondary and tertiary treatment plants. Primary treatment focused solely on the physical separation of dissolved solids from the water by sedimentation. WWTP's have large holding tanks which require a large footprint. In these tanks, the sewer water is allowed to rest, so heavier solids gravitate to the bottom where the sludge is siphoned out either for further treatment or, most commonly, to be landfilled. Secondary treatment plants have an additional step which is the sterilization of the effluent. In Mexico and countries which cannot afford expensive treatment options, at this point in the treatment, chlorine will be added to the water to kill off the pathogens and the solution will be discharged. Finally, and less commonplace, there is tertiary treatment or advanced treatment which is expensive and more common in large plants in rich countries. During tertiary treatment, the nutrients in the wastewater are removed. Nutrients such as Nitrogen, Potassium, Phosphates which are present in human excreta are not removed from the effluent otherwise. Recently, with increased urbanization and tighter government regulations, WWTP's are being forced to remove the pharmaceuticals from the effluent as well. However, once again, these advanced technologies are out of reach to most countries on the planet because they lack stringent environmental regulations, don't have incentives to curb pollution when they are resources-rich and there is little political will from traditionally corrupt governments.

At the time this was all I knew about wastewater and although I understood the concept, I couldn't motivate myself to study WWTP's. I didn't understand why I couldn't be interested in the subject. I've always been interested in everything related to Nature. Sports, after all, always put me outdoors. I remember going cycling at the Everglades National Park in South Florida. I've always cared deeply about the environment but for some reason, I didn't like ENV3001. The first time I took the course, I dropped the class because I didn't do well in the other topics. Then the second time, I risked it and got a D. Then the third time I got a C minus which was just short of passing. The fourth time I dropped it again. Finally, I took the course a fifth time and passed. Dr. Tansel who was the director of environmental engineering department lectured it. It was the last course I took, all by itself during my last semester at university and I was so worried about not passing it that I e-mailed the professor to make sure I had passed it.

I contacted Rafael in Spain again asking how the 2016 season selection process was going. He replied he wasn't done but he didn't believe I'd have a chance to swim in 2016 due to the quantity of swimmers waiting from previous years. Swimming the Strait of Gibraltar is a highly coveted athletic endeavor. Swimmers from around the world wish to connect Europe to Africa by arm strokes and kicks. Unfortunately, the swimming crossing season is only open from April to October. There can only be one crossing a day with one to four simmers at most and the swim attempts are dependent on the weather and the organizers will not schedule a swim when there are strong winds, tall waves or rain. Waves would add a significant distance to the swim. Would you rather swim in a flat line or in a sine wave? Each swimmer is given a five-day window to attempt the crossing. If the weather does not permit, then he or she must return the following year. These conditions made this crossing one of the most difficult to accomplish within the seven channels circuit and that's why Rafael couldn't promise me a slot.

Crossing the Strait was also a political move. Both the Spanish and Moroccan coast guards must authorize the crossings and they remain vigilant throughout the whole time. Innumerable people from drug smugglers to refugees have died while crossing these waters. The Strait might as well be a cemetery. There are also cargo ships the size and weight of small cities squeezing in the warm Strait waters while watchful eyes from both sides consent to their passage. Organization logistics and politics aside, I accepted that I needed to wait for my turn and continued to train, using my monthly triathlons in South Florida as training.

At the time, I had finished my internship at Odebrecht's Fort Lauderdale airport, and I arranged a transfer to the logistics department in Miami. I sought the move in order to be closer to my university campus as I finished my engineering degree. I also figured that it would be a good way to learn more about the company operations in different countries and then travel to places like Angola where Odebrecht had many jobs. I wanted to show them that I was mobile and ready to go anywhere in the world where there would be a construction job. However, this turned out to be a gross error on my part. Odebrecht was about to enter an incredible downfall and the logistics department was closed shortly after I came to Miami.

Unbeknownst to all of us working there, Odebrecht was the key player in one of the largest corruption scandals in the world. The graft scheme was uncovered through the Car-Wash Operation, *Operação Lava-Jato* in Portuguese. Brazilian federal police revealed how former president Lula, the most popular politician in the world, was using his political clout to gain the company construction projects in Brazil, Cuba and other countries in exchange for paybacks in real estate and hard cash. Odebrecht worked in Muammar Gadhafi's Libya, and other dictatorial countries. In Miami, which is heavily influenced by Cuban politics, Odebrecht USA, the independent subsidiary of Odebrecht, was facing political pushback which was keeping it from getting new contracts in South Florida. Furthermore, the international negative press didn't help the company.

I actually flew to Brazil in 2014 to interview with Odebrecht but they told me to apply with the Miami office instead. I did and eventually got in. During the long orientations, they taught us about their origin. The company had been founded by Norberto Odebrecht in Bahia, a state in the northeast coast of Brazil. He was the grandson of a German engineer who emigrated to Brazil in the 1850's. The company preached with religious fervor its philosophy detailed in three books called TEO or *Tecnologia Empresarial Odebrecht*. In the trilogy, they extolled the virtues of doing business correctly with ethics and all of those principles one would expect from a serious company. I bought into it, felt proud about this Brazilian company and wanted to be part of it.

Therefore, it was a huge disappointment when I saw Odebrecht disintegrating in front of my eyes. The CEO and the top executives went to jail and started cooperating with the police revealing how corruption with corrupt governments was the *modus operandi*. So much for TEO, I lamented. My plan of using politics and engineering to help others had naïve. Politicians and engineers were stealing money while the people they claimed to help never benefitted from the construction projects paid for with their taxes.

Luckily for me, I had left the company earlier before the bomb went off and the experience opened the door for me to work at another giant multinational construction conglomerate, French company Bouygues Travaux Publics. I became a project engineer at the Port of Miami Tunnel, a one-billion-dollar project consisting of building two tunnels from Watson Island under the Biscayne Bay waters onto Dodge Island where the port is located. This was the type of project which, at the time, I believed would help bring progress to Brazil and other developing countries.

Nine months later and the tunnel job was nearing completion. Florida Governor Rick Scott came down to Miami for the ceremony and cut the ribbon opening the tunnel to drivers and the occasional Critical Mass ride that went through the tunnel. Me and my Indian friend Manhmadh rented a Cadillac limousine, invited our co-workers and were amongst the first people to drive through the newly built tunnel. It was fun but I started looking for international projects which would allow me to travel do developing countries and work as an engineer. Sometime during my research, I remembered how I used to watch Formula 1 races on Sundays with my father. We watched Brazilian racer Ayrton Senna race his red and white McLaren in the rain, winning incredible races against French rival Alain Prost. It occurred to me that building a road is a lot like building a racetrack and I wondered what kind of companies design and build racetracks. I went online and found two, one English called Apex and another German called Tilke Engineering. I emailed my resumé to both companies and told them about my interest in working abroad as an engineer.

Upon visiting Tilke's website, I learned they were working on the renovation of the Mexican Racetrack called Hermanos Rodriguez. I must have watched Senna race in this track before it was taken out of the Grand Prix. Now, two decades later, México would be back on the circuit and work was about to start if not, must have already been underway. I mentioned the project to a Mexican colleague called Uriel who was working at the tunnel with me and when he saw the project on my computer screen, he turned to me and asked: "Why don't you go there and check them out?" I looked back at him and said: "You're right, why don't I go to México?"

I had always wanted to go to México, and this was the perfect reason to take me there. With my recently gained confidence to travel, I knew I could find a couch there. This was the end of December 2014 and I had already accepted a project engineer position starting on January 7th, 2015 with Plaza Construction. I still remember meeting the company recruiter named Anisa at a job fair in my university. I dressed up with my one and only suit, printed my resume and walked confidently to her desk. I had already looked up Plaza as they were snatching a lot of construction projects in Miami, their green logo banner was everywhere. The firm is originally from New York where they built a lot of skyscrapers. Miami, always growing, had plenty of works for them. I have done some standup comedy open mics and improv theater which have given me the confidence I needed at many circumstances like this one. We shook hands, she had a firm handshake. I looked at her in the eyes, talked about my work at Odebrecht's Airport and most recently at the Port of Miami tunnel and she invited me for an interview which I must have done well because they offered me a job paying $65k a year, with health insurance and bonus.

But why not go to México as Uriel asked me. Only God knew when I would have time off again. I called Anisa and asked her to begin work January 14th, one week later which would give me time to volunteer at the project and learn about this type of construction. She agreed to my request and on January 1st, 2015 at 7am, I kicked out my friends from my apartment who were still celebrating new year's at a party that I hosted and headed to MIA, Miami's airport .

I really enjoyed that apartment. I had been flat mates with Rei, my college buddy, and when he moved to New York City, he transferred the lease to my name. It had two bedrooms and two baths, was located in downtown, close to the metro trail and metro mover lines. It also had a long balcony from where I had a 180° of south Miami. To the west, there was the Dolphin Stadium, little Havana and Coral Gables. Directly south, I could see Downtown, Brickell, and Biscayne Bay. And to the east, I had a view of the Port of Miami with its Cruise ships and even the tip of South Beach. At first when I moved there, the rent was low thanks to all the homeless people doing drugs and hanging out in the empty parking lots which are common in most downtowns in the U.S.A. Overnight, when the homeless shelter closed, the homeless people were gone and the rents around us starting increasing. I started missing the homeless people and drug addicts on the streets. Luckily, my rent didn't increase as much as the other buildings and I was living in one of the cheapest buildings in downtown at the time, only $1,200 a month rent.

I didn't know anyone in México, but it didn't matter because I had couchsurfing. Mexicans were extremely hospitable and all the requests I sent were accepted. It didn't matter that hours before my flight, my host cancelled mentioning some last-minute change. All I had to do was to send a message to my next host, Flores, and tell him I was on my way and needed a place to crash. He informed me he was already hosting other travelers, but I could still surf on the couch. No worries. I flew in and took the tram to the outskirts of smog filled México City. His house was very simple, and not well kept but it showed me that Mexicans were incredibly humble.

When I arrived in México City, I immediately felt at home. I don't know if it was because I had seen it so many times in *El Chavo del Ocho*, a popular Mexican TV that plays in Brazil, or if it was because it felt a lot like my home-country. Perhaps, it was the fact I lived in Miami and it prepared me to live in a Spanish speaking city. It may have been the unparalleled hospitality with which I was welcomed. I believe my experience hosting travelers allowed me to be a better traveler as well, helping my hosts and being open to changes. When I got to Flor's house, I met his other surfers and we went out to the grocery store to prepare dinner. Couchsurfers know better to save money by cooking with their hosts when they can.

The next day, I got ready for work. I put on my Bouygues hard hat, reflective vest, boots and took the tram and metro back to the city's racetrack. It was January 2nd, a Friday and I wondered if anybody would be working. Living in the U.S. I heard mostly negative things about Mexicans, and I wondered about their work ethic. I walked into the racetrack which sits inside a beautiful city park and started walking along the track where Ayrton Senna sped at lightning speeds. Some sections were cordoned off with signs of demolition work but apart from the resident's rollerblading and cycling, the track felt sleepy.

I continued walking and was surprised to meet a construction crew working on an excavator with a hammer attachment demolishing the old pavement and side curbs. I told them I was an engineer and that I wanted to volunteer at the construction for the next two weeks. They looked puzzled as I'm sure they don't get that type of question often. However, they told me to come back on Monday and speak with the *Ingeniero*, project superintendent. It was almost lunchtime and the surveyor crew invited me for lunch. Surveyors are those workers who are always looking into machines set upon tripods in the middle of the road or construction field. They take measurements and tell us where to build or dig. Their lines, like the swimming lines, are crucial to the project. The company was headed by Señor Salvador and his son Éder was second in command. Upon meeting me and hearing my story, they invited me to celebrate their matriarch's birthday in their hometown Tepatepec which would be a two-hour drive from México City. I felt like that was a unique opportunity and my couchsurfing spirit couldn't say no to such a singular experience and I agreed to go to their hometown. But first I had to rush out and get my bag at Flore's house and then run back to the track before the workday's end. It was quite the adventure given México city's chaotic traffic and the fact that my phone was about to die minutes before they picked me up at the bus station outside the airport which was close to the racetrack. I packed in the bed of the truck under the fiber glass canopy, next to shovels, and surveying equipment. It was already getting dark and I was so tired from the day that I fell asleep and woke up in a sleepy village breathing clean air and feeling like I was home again.

The weekend was filled with eating the most amazing food of my life. *Barbacoa de Borrego*, they call it. Borrego is their word for goat and barbacoa is a traditional barbecue method where they use volcanic rocks to cook meats in the ground over many hours, even days. The result is the softest meat one can wish for. I was so glad I wasn't vegetarian back then. There were many types of meat, cow brains and tongue; no piece was wasted. The catering company sent their staff to prepare and cut open the *gorditas*, special maize tortillas which inflate when they're heated on a flat iron. There were many spicy sauces which frankly speaking made anything taste great. The drinks were Coronitas which I didn't know until then, basic coronas in tiny bottles, baby size. Nonalcoholic beverages consisted of Água de Horchata and de Jamaica, rice and Jamaica flower juices respectively. I met the whole family. They treated me like a cousin from the north. I hugged all the *tías* and *abuelas*, met *las primas* and sang drunk mariachi songs late into the morning with my new family.

On Monday we came back into the city and I went in for my

interview with the *Ingeniero*. I sat in for a proper interview and he allowed me to volunteer on the site and joked that I'd get paid one taco per day. I took a look at the drawings from Tilke, read the manual detailing how the new pit would be constructed, met two of their German engineers and worked with the survey team during the two weeks I spent in *Ciudad de México* or CDMX. In particular, I remember using chalk powder to draw a line where a new track portion would be built through the old baseball stadium which was inside the Hermanos Rodríguez park.

I had to say goodbye to my new friends and family, but the experience profoundly marked me. I learned that I could create my own opportunities and that I didn't have to wait for a company to take me anywhere. I e-mailed both English and German companies telling them I had volunteered at the track and hoped they would get back to me.

3 – SHALLOW

In January of 2015, I started working at Plaza. I was placed in a jobsite nearing its completion. It was a residential building in Brickell, a fancy area south of downtown Miami. The building had 60 floors and about 400 units. This part of the project is one of the most stressful. The project managers were busy trying to pass the fire safety tests with the fire department in order to meet the deadlines on the contract. Every week, we had new tests and inspections by the local authorities to make sure the building was safe for occupancy before we could turn over the keys to the owners. I remember walking the floors and stairs checking which unit was completed and what needed to be done in the ones unfinished. It was tedious work, but it was critical to the project completion. At one point, I was handed all the keys to the entire building and had to make sure they all worked. That was a lot of fun.

During this time, I decided that I would learn French that year. In fact, since 2008, when I met my friend Arnaud in a semester abroad in Buenos Aires, I told myself that I would learn French. Also, my Couchsurfers had inspired me to travel and I had planned a trip to Europe that summer. I never really liked taking language courses, as I preferred learning on my own like my friend Efra. So instead, I downloaded a mobile app called Duolingo which teaches foreign languages through a gamified strategy. Their motto is to make language learning free and fun and it worked for me. Whenever I had any downtime, instead of wasting time on social media, I would play the French Duolingo course. A lot of my job included waiting for subcontractors to come to the unit where I was and fix something and whenever I could I played the game. Duolingo has a points system which allows you to compare your progress with others and a healthy rivalry develops between learners.

And when I realized that most of my couchsurfes were French, I decided to speak with them. I was shy at first and all I could say was some words from the classic "Ne me quitte pas" song by Jacques Brell famously played in a Brazilian series called "Presença de Anita" which dominated my pre-pubescent youth. Initially, the conversations were brief. "Çá vá?" which is "Hello, how are you?" We would start in French and I'd go as far as I could before we switched back to English. I found many French to be self-conscious of their accents. They love to say: "My English is shit." Personally, I love accents. They tell the story of who we are, and I find accent reduction courses to be a waste of effort and money. I took advantage of this embarrassment and pressed on speaking French. At first it was hard because they always wanted to switch back to English when they noticed me struggling. However, I kept on studying through the app and practicing with them. I realized that my French started improving dramatically once I learned how to ask questions. To me, the two most important questions were "Çá veux dire?" and "Comment dire x" which respectively mean "That means?" and "How do you say x?" note that these are informal versions of the questions, but that's how my surfers were speaking to me and that's how it felt most natural. Gradually the conversations got longer in French. I started picking up on slang and most common expressions and within four months I was speaking at a basic conversational level. My progress surprised my French surfers and myself equally. I couldn't believe I was learning French without a textbook, a professor and in my living room in Miami.

When I speak a foreign language, it's as if I'm high. It's a great feeling to hear myself in another language, understand others and be understood. It satiates a need that I have to connect. I love the feeling that I get when a foreigner hears me speak in their native tongue. They look at me with different eyes, surprised and eager to share more about themselves. We do tend to speak more openly, from the heart even, in our native tongues. It seems that there is a stronger bond when we speak the same language.

I had swim coach named Lisa who drilled into me the importance of the hand entering the water. Every time, I entered the water too early, I was cutting myself short, undermining myself, shortening my distance swam per stroke. Don't come in too early, was the lesson.

Another lesson came from another swim coach named Abbie Fish. She suggests thinking about the arm stroke as about doing a bench press exercise or doing pus-ups. The farther your hands are apart, the harder it would be to bench press or push yourself up. However, in swimming you don't want your hands to go into the water too close to your body. As per Abi, there are three steps to the arm stroke cycle:

1.Entry and extension when you put your hand in and stretch your arms

2.The catch which is literally catching water and

3. the pull which is when you pull the water that you caught back propelling yourself forward (thanks to Newton's Third Law of Motion).

Whenever, I am feeling slow in the water I remember Lisa and Abbie's instructions to correct my swimming posture and I can feel that I become more efficient in my efforts. In order to swim long distances, one must save energy and each arm stroke counts.

There's nothing quite like the feeling that I have when I climb out of the pool after a long and hard swimming session. It feels great when I leave everything in the pool, meaning I gave my best effort and I have no energy left. Practices tend to last one hour to one and a half. After I push myself to the limit, I feel at peace. I do a push up to get out of the pool, remove my latex cap which was compressing my head, take out my goggles and earplugs as I walk to the locker room. I feel unparallel peace as I my bare feet touch the pool deck. There are no thoughts in my mind, and I can't talk to my chatty teammates. I don't have any energy left. I feel like I accomplished my mission and I start looking forward to food to fuel me up, restore my broken muscle tissues and a nap.

One day I got an e-mail from Clive at Apex, the English company designing Formula 1 circuits which I had contacted prior and after going to México. He was replying to my e-mail which I sent after having volunteered at the racetrack in México City. He liked my initiative and invited me to his office outside of London. I informed him of my European trip that summer and we agreed to meet then.

I continued reading. In fact, I read a lot. Perhaps it has to do with the fact that my father used to sell encyclopedias from door to door in Brazil. Growing up we had our own Barsa encyclopedia which was like having our own Wikipedia before the internet. When I asked my father a question, he would often refer me to the encyclopedia. I learned that I could learn what I didn't know by reading books about different things. Most importantly, I realized that I could obtain knowledge. I didn't have to wait until an adult at school chose what topic he or she would teach me.

When I was thirteen my parents moved me to a better school in Sobradinho. It was the best school in the city, private and catholic. I was at an odd situation because most kids were there because they were from well off families but the only reason my parents could afford that school was because my mother was cleaning houses in the United States and sending remittances to my father to pay for my school fees and my sister's. With my mom's help, my father could afford to pay for my other three siblings' fees. We were five in total and those fees added up

But I could tell some of the other kids were also in similar situations, with parents struggling to put them in a good school and ensure a promising future. This economic disparity between me and most of my new classmates and the fact of changing schools after almost ten years studying at the same place, where I knew everyone and everyone knew me, was a seismic shock to my social circle. Overnight, I had basically no friends apart from the pastor's older daughter from the church I used to attend. I didn't have anyone to play or talk with during lunch recess, so I'd eat my *esfiha de frango* (a chicken or ham pastry which to date is the best I've ever had) in a sunny spot of the concrete patio. It felt a bit like the sun break prisoners are allowed per day. And since I didn't have much to do, I'd go to the hallway and wait for the bell to ring while standing in front of my classroom. One day the *Orientadora*, which was the school advisor, asked me why I was waiting for class by myself and I told her I didn't know anyone. She gave me a reprimanding look and told me to do something else, in a way only a catholic nun could and walked away in her white robe.

Since I wasn't allowed to be alone, I decided to walk to the end of the hallway where there was a library. It was minimalistic. There were a couple of grey, thin steel and sparsely populated bookshelves with equally thin books lining up the perimeter of the room. Two tables populated the area and the librarian's desk was in the right corner when you walked in. That library represented freedom from my barbed wire , private school , catholic prison. After a few months I read all the books which interested me and ran out of books to check out.

Before traveling I read Elon Musk's biography. He was born in South Africa to a model mother and an engineer father. When he was a teenager, he decided to emigrate to the United States because he wanted to be part of the internet revolution going in in Silicon Valley. I also read there he thought that Burning Man, that festival about which my Couchsurfer Rasa had vaguely told me, was Silicon Valley. I liked reading his bio because it described how he employs first principle's reasoning which means to think about a problem down to its atomic level. For example, Elon Musk is well known for his SpaceX company which builds rockets that are servicing the International Space Station (ISS) and is soon to become the first American private company to take astronauts to the ISS since the shuttle program. It was the smoke trail of a SpaceX rocket which I witnessed launching from Cape Canaveral all the way down in my running track on a dark morning in Miami.

SpaceX is successful for among other reasons, the fact that it recycles its rockets. Previously, NASA and private companies would build rockets to take a payload to space and upon reaching the desired orbit, the rockets would fall and crash in the Atlantic Ocean. Elon Musk realized that in order to have a competitive rocket company he needed to land his rockets back on the launching pad, similar to how airplanes work. Can you imagine how much plane tickets would cost if, after every flight, the airline had to buy a new plane?

With this mission in mind, he built SpaceX and designed the Falcon rocket which lands back on the same launching pad from which it launched. In order to reach this goal, the engineers decided to use grid fins, previously employed in missiles, which control the flight path of the rocket back to the launch pad where prior to reaching the ground, the rocket engines fire again and propulsively land the rocket. I invite you to watch the launches and landings on YouTube, it's an amazing feat of engineering.

As an engineer, I appreciated this thinking. Come up with the constraints, in this case make a rocket that lands itself and get it done. And that's how I'm approaching the toilet of the future challenge. While SpaceX calls it "rapid reusability," what they really mean is recycling one's resources. This thinking is exactly what I am applying to the toilet of the future which I'm building. It makes business and environmental sense to wisely use one's resources, be it rockets, water and even "human waste."

During those early Try 2-1 track trainings, I developed a friendship and rivalry with teammate Roberto who was three years younger and a lot faster than me. We had competed in the same races before and I wanted to finish 2015 by beating him. Early on the year, I signed up for the 2015 Key West Sprint Triathlon. Since I had intensified my training, thinking of Gibraltar, I started feeling some knee pain and went to the doctor who diligently told me to stop running. He may have said something about me not being fit for running but I don't remember exactly what he said. Instead I looked at alternatives to supplement my training. Being a water person, it intuitively occurred to me of reducing my weight by running in a pool with water depth up to my chest. Water, being denser than air, also forced me to correct and improve my running posture. At the same time, living in a tall building in Miami, I thought of running upstairs. When running upstairs, I felt less loading on my knees while doing a stronger cardiovascular work-out. During the first attempt, I could only run up three floors, then I gradually increased until being able to run the whole building, 30 floors one month later. This was so intense that I almost vomited or passed out a couple of times. I had a heart rate monitor and the line peaked fast, going off the chart during the first minute. I eventually managed to run up thirty floors in less than three minutes. It was also around this time that I tried an ice cream diet challenge. I'd eat ice cream in the morning before workouts and still managed to lose weight. It was such a great motivation to eat chocolate ice cream each morning before my workouts. You should try it.

Summer came and I went to France where I tested my language skills. I missed a flights in Paris and had took another one by speaking with an attendant at the airport and I got away with my basic Duolingo French. I got into a virtuous cycle where it was easier and pleasant to keep on learning by speaking with more people. I had also read that there are about three thousand words which are spoken 95% of the time. In the Duolingo website you can see how many words you know and at the time of this writing I only learned 1,500 words and most of them I already knew given their Latin origin. However, learning French became even more fun once I realized that it was not a task with and end-date. There will never be a day when I will say: "That's it, I speak French, no more studying for me." Learning is a life-long activity. If I stop using the language, then I will forget it and the way to learn is by continuing to speak it.

During my European trip, I visited Apex and spoke with Clive about working with him. He was impressed with how I went to México and invited me to attend the FIA conference that would be taking place in México city in a couple of weeks. FIA stands for *Federartion Internationale del'Automobile* and it's the governing body of Formula 1 and many auto races across the globe. The name is in French because they were founded in Paris. Once I realized this, I remembered a famous video where Ayrton Senna fights back against one of FIA's executive, Jean-Marie Balestre. In the exchange, one cannot help but hate the Balestre's French accent, especially if you side with Senna. I obviously said yes to Clive as this would allow me to see more of the F1 world, so I got two days off work and went to México, courtesy of Apex. Is this what I had dreamt? I wondered. Formula 1, to me, is a lot like a circus for the rich. These teams travel around the world in one year, competing in races every two weeks while lavishing out expensive brands.

I also visited Germany. I went Aachen, a university town, where Tilke Engineering was based. I met with a project manager and them about my México trip, but unlike Clive, they were not impressed. I toured their office and went to Manheim to meet my friend Walter who I met while swimming with Nadadores in Miami. He welcomed me with euphoric enthusiasm, open arms and a big hug I wouldn't expect from a German. He split his time between swimming in Miami and his patients near a small city called Pirmasens. I met his family and later his daughter took me to a local bar to meet her friends and I was impressed with the fact they could still smoke indoors. It was a tiny village.

I flew back to Miami and decided to take care of my university obligations. I had finished my engineering studies at FIU but in order to graduate with a diploma in Civil Engineering, I had to first pass the Fundamentals of Engineering Exam, or FE Exam as we dearly call it. The exam is the first in a two-exam path to become a Professional Engineer in the United States. It consists of 100 questions from the four years of engineering and we have only five hours and twenty minutes to answer all of them. That means, we have 3.2 minutes per question. It's administered by a private third-party examination board and the exams are computer based, that is you sit in front of a computer and choose from four multiple questions. They have a strict calculator policy and tight, airport-like security at the testing centers.

However, it's hard to run one's brain for this long while doing higher level math and complex engineering calculations. My university offered a preparatory class, but I was confident of my ability to pass it by studying on my own with a heavy preparatory book and practice tests. Weeks later, I sat in for the exam and walked out of the testing center mentally exhausted and sure that I had failed it. Then, I got the unsurprising result that I had failed the FE exam. I was not upset, however. I was starting to doubt everything I had studied and learned until then. I was feeling lost about what I was going to do with my life. I didn't know if I wanted to become a professional engineer and I knew I didn't want to be in the United States for much longer.

I was working as a project engineer for a fine company and getting paid well for my age and experience. I had attained the American dream by any definition of this term. I was living a comfortable life in Miami, traveling to Europe and having fun. Professionally, I was on the path to become a project manager, but I wasn't feeling it. There was something missing. I didn't know what, but I didn't feel true to myself. I wanted to explore the world, use my skills to improve the lives of others. I wanted to use engineering to solve problems and building another high rise in Miami was not enticing me anymore.

June came around and it was time to attend the F1 conference in México City. I only needed to take two days off, Thursday and Friday and Plaza didn't need to know about it. I flew to CDMX, and this time I didn't need to ask for a couch because Apex was paying for everything. Was this what I had dreamt of? I asked myself with a big smile on my face. México was treating me well. The conference consisted of listening to big Whigs in the F1 world with their French accents, and Carlos Slim's Son, showed up as well. He was the son of the richest man on the planet at the time. In my opinion, Mr. Slim was rich because he enjoyed a monopoly with his *Telefonica* mobile service provider and his buddies in government. I learned about Apex's jobs around the world including F1 tracks in Dubai, Malaysia and other countries. I talked to attendees about potential racetrack projects Central America and Africa. It was exciting but Apex had not gotten any new work which would warrant hiring me. At the end of the conference, we were invited to a gala dinner in a monastery inside of México City which I didn't even know existed. Super models were in charge of taking our pictures and checking our credentials. Although, I hadn't gotten an offer, I left the México with a badge from this conference; it had my name, the FIA logo and the date June 2015. This badge would prove to be useful at a later date.

I went back to Miami and was transferred to a jobsite in South Beach, a luxury residential building. The type of real state that the cynics among us suspect will be used to launder money from Russian and south American drug lords. Miami can be a bubble. I'm embarrassed to admit I lived there for so long. To be honest, it didn't feel like I was living in the United States. I mean, learned Spanish and French while living in South Florida. I had more contact with foreigners than native-born Americans. In order to experience the real USA, I had to drive a couple of hours north into Orland and it was scary, so I limited my trips that way.

October approached and back in México, the racetrack had been completed and the race would take place on November 1st, during the day of the dead, *Día de los Muertos*, celebration. Mexican culture treats death in a very special manner. It's a three-day, if not week-long in some cities, celebration of those who passed. Altars are placed in living rooms with pictures, and favorite foods and drinks of the deceased. While researching the Aztecs, I learned that they believed that dogs were our guides into the afterlife. Once one died, his or her dog would be put down and buried with its owner so that the dog could lead them out of darkness in the afterlife. I enjoyed the thought of meeting my dogs from my childhood. How bad could death be if we could still hang out with our pets?

I appreciated the beliefs surrounding death, but I was more interested in attending the race. After all, I had volunteered early in January, attended the FIA conference in June and now it was race time. However, tickets for F1 races easily reach five hundred dollars and although I could afford it, it would take a huge chunk of my salary. I was saving everything I could, so I decided to buy the plane tickets to Mexico and try to get in the race somehow because I knew the track inside and out. It was surreal to board the plane, arrive at my friend Paulina who I also met through couchsurfing back in January, and wake up on November 1st, shower, brush my teeth and go to the race for which I didn't have a ticket.

Yet, that is exactly, what I did. I packed my passport, my FIA badge from June and boarded the metro to the Hermanos Rodriguez metro stop. The metro was full of ticket holding F1 fans and I was just wondering how I would watch this race. Worst case scenario, I could watch it from a bar in front of the racetrack. I just wanted to enjoy that day, I realized. I got out of the station and started walking around the perimeter of the track which was full of fans trying to find their gate and there were cops everywhere. I quickly threw out the option of jumping the three-meter tall, barbed wire walls and continued walking around the track.

Until one police officer, standing in front of a small blue door looked at me and asked me in Spanish: "¿Éres Prensa?" meaning, are you with the press? I lied, saying "Sí" and he invited me inside the press area which consisted of an auditorium with a large flat screen showing the racetrack and the drivers warming up the engines. I was in shock but not satisfied. I was so close, I could hear the engines louder than the TV, so I went outside and spotted a green tent with barricades around it.

I walked up to it and saw two young guys wearing military uniforms checking credentials on a laptop. I walked confidently up to them and handed in my passport and that FIA badge with my name on it and the date "June 2015." They proceeded to check my name in their laptop but obviously didn't find anything. I told them I was with FIA, lie #2, and they looked at each other, shrugged their shoulders and wrapped two bracelets around my wrist, allowing me to go up into a building which I didn't remember existed. I went up the five flights of stairs and realized those bracelets were my entry into a VIP-area with a lot of alcohol and a banquet. That was my first F1 race experience.

It got better after the raced ended. I walked on to the track, through the baseball stadium area which I had helped draw the pavement contour with chalk lines. I wished Eder and Salvador could have been there. Then I said why not go to the other VIP area, the paddocks where the race drivers hang out with celebrities and talk to reporters. I continued walking across the newly built track to the new garages and I met a racing legend who has recently passed away, Austrian pilot Niki Lauda. I also met Lewis Hamilton and Sebastian Vettel. Even though I stopped watching F1 a long time ago, it was still great to meet those drivers in person.

And to finish this amazing day, I met up with my Mexican friend Martin Cortina who had volunteered as a race marshal which are those guys and girls wearing high visibility orange vests and stay positioned at keys points throughout the track and will assist drivers in case of accidents. I met Martin and his cousin, Anna once when they stayed at my apartment in Miami, not through couchsurfing but Airbnb. One roommate of mine had left suddenly and I had to cover one month's rent that way. Martin and his Marshalls friends got together were having their own fiesta inside a temporary tent. They had Mariachi music blasting and a conga line which had a mandatory stop where one of the marshals was pouring Tequila into the dancers' mouths.

I got back to Florida and after six months of training, it came time to compete in the Key West Triathlon. I told Roberto I was looking forward to racing against him and I was happy to know he had actually signed up for the race. I drove to the Keys with my mother and swimming buddy Alberto, an Italian guy who had been shipped off to Miami by an organic pasta-making company. I got a couple hours of sleep and on Saturday morning, lined up at the beach to start the race. There are two types of starts in triathlon, dry and wet starts. As you can imagine, a dry start is on land. The horn goes off and everyone runs into the water and starts the swim; I was mostly used to this type of start. However, this time, the race organizer Multirace decided to change to a wet start which is how Ironman Triathlons begin. Athletes go into the water, acclimate themselves to the temperature and egg beat like water polo players until the horn goes off. It was my first wet start, but I wasn't more nervous than usual. Since I'm a strong swimmer, I like to be in the front and to the right to allow for any faster swimmers to pass me on the left. I learned to be in this position because when I started and was much slower, I'd get hit by fists and elbows of swimmers coming behind me.

I got to the front of the pack and waded in the water, heart pumping, waiting...until the air horn blasted. I started swimming but one guy behind me clawed off my timing chip, a Velcro anklet with a radio transmitter which remotely lets race organizers track each time we walk over receiver mats placed at the water edge, transition zone and finish line. This had never happened to me. In fact, I learned a hack from pro-triathletes who added an additional layer of security to the anklet by taping it over with black electric tape. But on this day after years of competing, I failed to tape my anklet and for the first time, somebody took it off me. All of this happened in milliseconds. I felt the anklet slipping off and there was nothing I could do. The start of a triathlon is one of the most adrenaline-laden moments of the sport and I would never be able to swim back against the swimmers' wave coming hot on my heel…or ankle. And even if I did, the timing chip would probably have sunk, and the turbid waters would have made it impossible to find it. I could only swim forward and that's what I did.

I thought of all my training during the year, my stair climbs, water sessions and wondered if it had all been in vain. These self-sabotaging thoughts were creeping in my head while I swam as fast as I could. I also figured I was already in the race, in Key West and might as well enjoy it, so I continued swimming. When I stepped ashore, I ran towards the transition zone which is a rectangular area where we keep our gear and bikes in racks. When we change from swim to bike, we call it, transition 1 or T1. The transition area is fenced off to keep the bikes and belongings safe. There are only two entrances on each end of the zone and when we run in, we are forced to walk over a mat which picks up the radio frequency of the anklet and records the time each athlete took to swim.

Since I didn't have mine, I frantically looked around for race marshals or volunteers to quickly let them know I had lost my timing chip. As I ran into T1, I spotted a Multirace volunteer sitting down by the entrance and I managed to shout that I had lost my chip and yelled out my race number "**453**." However, since I was running the volunteer didn't have time to register my message and looked more startled than helpful. I thought to myself "whatever" and dropped my swimming gear, googles and cap, grabbed my helmet, bike and started running out of the transition to start the bike portion.

As I pedaled, I had 35 minutes to think about dropping out of the race and wondered whether I would be able to claim my finish time. Again, I pushed off these negative thoughts and cycled for 20 kilometers. It came time to transition into the run, T2 and I again looked for a volunteer and this time made sure to point to my ankle and yell "**453**." Again, I left another volunteer looking like a deer in the headlights. I dropped my bike, helmet, slipped in my running shoes and started running as fast as I could for the next 5 kilometers or 3.2 miles. One minute after turning around the halfway point, I spotted Roberto for the first time in the race. Seeing the lead, I had over him gave me a confidence boost and I ran faster still. I didn't want to lose the race to him at the last meters, as something similar had happened to me earlier in the year, so I dug deep and ran faster than ever.

Once I spotted the finish line 500 meters away, I looked back to make sure no one would steal my place and upon realizing I was alone, ran with a smile on my face. I crossed the finish line, caught a lung full of air along with my medal, and told a race official that I had lost my timing chip. The Multirace staff assuredly patted me on the shoulder and pointed to a lady sitting by the finish line writing down the numbers of each athlete that crossed it. They had a manual backup system in case of any technological failure. I was surprised and relieved at the same time. Minutes later, I walked to their mobile timing unit to print out my timing slip and found out I had run my fastest 5k ever, 4:21 mins/km (7:06 min/mile). It had been worth it. I beat my teammate, reached my personal goal for the year and reinforced the valuable lesson I had learned in Key West. I didn't give up.

4- JUMPING OUT

When I started furnishing my rented apartment in Miami after Rei left for NYC, the first piece of furniture I bought was an Ikea bookshelf. It was two meters long or six feet and it took half of my living room wall. I put all the books I had read and continued buying over the years. One such book *was The Humanure Handbook* by Joe Jenkins. Joe is a roofer by trade, but he literally wrote the book on compost toilets. He advocates a simple bucket system and shows how anyone can have a compost toilet at home for less than $25 dollars. I highly recommend his book to anyone interested in compost toilets and ecological sanitation.

Ecological sanitation is the concept that human wastes are really not wastes at all. When one elephant takes a pachidermic dump, it doesn't contaminate the savannah where it lives. Instead, its dung becomes a valuable resource for insects. Dung beetles for example, form dung into balls and roll them into caves they've dug, enriching the soil by bringing those nutrients underground. Feces, including our own, contain nutrients from the food we ingested. Pigs only extract 30% of the protein they consume. Similarly, our own bodily excrements still contain much of the nutrients which were up taken by the plants and made up the meats we consumed. What we eat is what we excrete.

Ecological sanitation also teaches us about the nutrients cycle.

When we grow food, we must first fertilize the soil with nutrient fertilizers such as Nitrogen, Potassium and Phosphates, (N,P,K as per their element symbol). The fruits, grains and vegetables we harvest uptake these nutrients from the ground and we subsequently consume them. Afterwards, our bodies digest the food, absorb the water, vitamins and nutrients and excrete what was not used for our nourishment along with bacteria that was lining our stomachs and undigested fibers.

Nonetheless, what may be surprising to many is that our feces and urine still contain, besides nutrients, a lot of energy. To put into perspective, dried feces have the same calorific value as wood or 17 Megajoules per kilogram. That means if we dehydrate feces from its moisture content of 75% down to 30%, then we can make a fire fueled by dried poop just as hot as if we were using firewood. We can also produce methane gas through anaerobic digestion. But even more practically, human waste can be composted and turned into fertile soil which can grow food again. If you stop to think about it, that's how things have always been on the planet earth. One animal's "waste" is another's "resource."

If you look at shit under a microscope, you will realize that it is not waste, but a precious resource. And if you eat a healthy diet, rich in protein and nutrients, then your shit will be rich in nutrients as well. Just because it's being expelled from our bodies, doesn't make it any less valuable to the cycle of nutrients. We live in a closed system which has many cycles, such as the water, nitrogen and carbon cycles. Shit is not a waste, I've learned, and I hope to inculcate this into your head. Going back to the microscope, we will also find bacteria in our excrements. These bacteria present in our excrements came from our guts; they helped us digest the food and once out of us, they will continue digesting and decomposing our shit until it becomes composted soil ready to nourish new food, continuing the natural cycle.

Sadly, we've forgotten about this cycle and we've replaced it with an unnatural line where food is taken at one point, sewer is created by defecating in potable water and wastewater is discharged elsewhere, creating pollution for someone downstream deal with at a high energy cost. In my opinion, the biggest culprit of this nutrient cycle breakdown is the water-flushing toilet. It took me a while to internalize this, however.

Until its introduction halfway into the 19th century, all civilizations from every corner of the globe had managed human waste ecologically. That is collecting, composting, and returning the nutrients in our "wastes" back to the soil from which they came. When water was made available at the house, with the turn of a tap and the water closet got its own pipe, then suddenly, a lot of water was being used to wash away porcelain bowls with shit. Back then, there was no plumbing code or regulations and the easiest way to deal with sewer was to pipe it to the local river valley or discharge it into the ocean. Water, after all, always finds the lowest point. But, back then, there were no treatment plants. There was no scientific agreement that diseases such as Cholera were being transmitted from human feces in contact with water. The little and prevailing knowledge was that "dissolving was the solution." That is, people thought that once sewer was dissolved into the river, then there were no problems. The Germ Theory of Disease, that is the consensus that diseases were spread by microorganisms didn't become accepted until 1860's.

And even after the scientific community learned about viruses and bacteria, it took a long time for the wastewater treatment plants to be built. Major cities like Paris, New York didn't get treatment for another century. I love to mention the fact that the sewer generated by U.S. presidents in the White House up to Richard Nixon in 1974 didn't get treated. In fact, two sitting U.S. presidents have died from waterborne diseases, Willian Henry Harrison and Zachary Taylor. They're believed to have died from Typhoid Fever and cholera or diarrhea, diseases common in "third world countries."

I cringe at this term. As far as I've seen, there is only one world which we all share. Whatever happened to "Second World Countries?" This term perpetuates ignorance and encourage a distance between us and them. How many Americans know that their very presidents have died from diseases which are still common around poor countries today? As I've said before, we assume that the wastewater is being treated, and that it's been treated before, but wastewater treatment is a novel approach to human excreta and, in my opinion, an unsustainable one. If even wealthy nations took centuries to build theirs, then how can we expect struggling economies to catch up in a matter of years?

I believe that as the planet's population grows and our need to better manage our resources intensifies, we will be forced to reconsider how we look at shit; it's only a matter of time. You may have heard the sentence "water, is as our most precious resource," or "we are 60% water," and we can't live more than three days without it. Water is also essential to irrigating the food we need to survive. Yet, if water is so important, then why do we shit in it?

Our modern infrastructure treats water as if it was worthless. We're teaching ourselves that it's okay to pollute water because we are advanced civilizations and we can clean it later. Worse yet, our modern sanitation system misleads us into thinking that modern toilets should pollute water. But do we drink the water that comes out of a wastewater treatment plant? No, we don't because it's not fit for drinking. It still has nutrients from our shit and piss, as well as chemicals used to "treat" it, notably Chlorine, the same chemical used in the pools I train. If the effluent coming out of wastewater treatment plants had to be potable, that is drinkable, then these plants would be even more expensive to build, run and maintain, in fact prohibitively expensive. Therefore, by definition, our *modern* water infrastructure is designed to pollute potable water.

Consequently, our contemporary habits have removed us from

nature and we no longer know our place in it. Since we no longer need to take care of our own shit, we have forgotten that if safely composted, it can be returned to the soil as a fertilizer. Instead, we flush and forget, not realizing or knowing that, downstream, our shit is impacting the waterways and eventually ourselves. We have forgotten that we are nature too. Strikingly, the modern household toilet is the leading water user in the home, accounting for 25% to 33% of water consumption. Think about this every time you flush once after peeing and twice when you take a shit, because who flushes only once after taking going number two? I know, I don't. On average, humans urinate 2 liters and defecate 150 grams per day. If all water used in the home is used to flush an average two liters of urine and 150 grams of feces (5.3 ounces). That is more than 50 liters are used to flush away less than three liters of nutrient rich human waste. If we apply Elon Musk's first principle thinking to this problem, I'm sure we would use a different type of toilet.

Ironically, when we experience a natural disaster, we temporarily come to our senses. In 2005, I lived through a hurricane in South Florida. The supermarket water bottle shelves were wiped clean as everyone stockpiled water jugs at home, bracing for days without water service. When the hurricane reached land, powerful winds knocked down trees on powerlines and we were left without electricity and told to ration any available water. We didn't flush the toilet after we peed and didn't use bottled water to flush the toilet. Luckily, power lines were repaired days later, and we didn't run out of water, but the warning lesson remained.

Is this what it will take for us to change our toilet system? Are we going to wait until the system collapses? As countries grow, populations become more urbanized, further stressing the water supplies. Informal settlements, slums or *favelas*, are a real challenge to every growing economy around the world and these citizens want to flush their shit away just as much as you do. Water used for flushing the toilet is the low-hanging fruit in the residential water consumption matrix. A dry toilet is a solution which allows virtually anyone who is willing, at little expense and in no time, to take care of their own shit. But this takes civility, education and conscience.

Disturbingly, the notion that water flushing toilets are – or should be – the standard is misleading. According to the World Health Organization, two thirds of the world's population do not have access to sanitation once you consider wastewater treatment into the equation. That is, some people may even have a nice white toilet that is flushing raw feces into the local river or ocean, but no one is noticing it. This is the white lie of water flushing toilets. Also, according to the WHO, two thousand people die every day from diarrheal diseases, mostly infants and the elderly—those with vulnerable immune systems—because they are drinking water with shit in it. Guess where the shit is coming from? Their neighbors' and their own shit contaminating the local water supply is coming back to kill them.

All of this information was swirling in my brain as I tried to work in Miami. I had recently been moved from the jobsite in South Beach into the main office, only a 15-minute bike ride away from my apartment. Previously, a tourist had crashed into my car and I had been happily riding my bicycle to work every day. I felt quite blessed to be the only person in the office having the luxury of living so close to work and brave to face Miami drivers. Not owning a car allowed me to save serious money by saving on insurance and fuel. I used my savings to invest on Tesla stocks (another company started by Elon Musk) and fill up my savings account, something I had not been able to do before.

Even though I was happy to ride to work and felt comfortable at my job, I continued feeling that I was doing something wrong. My performance started slipping and I couldn't get excited about my work. Since I had been transferred to the office, I had been tasked with administrative work which did not appeal to me at all. There are a lot of lawsuits and legal proceedings in U.S.

construction. Mostly, subcontractors were suing the company or being sued by us and we needed to negotiate deals or settle out of court to avoid expensive lawyers. I remembered one warning from Rodrigo, a project manager that I worked with at Odebrecht who told me: "the higher up I go in the project ladder, the more I feel like a lawyer." I guessed I was moving up.

I was asked to work on two such cases and my routine had been reduced to talking with lawyers, consultants and doing a lot of paperwork. I was not enjoying coming to work anymore. The best part about work was my bike ride to and from it. One day, while looking up details of a jobsite, I noticed that water flushing toilets in luxury buildings in South Beach were being installed at the cost of $2,000USD each. Ten blocks north of my job, new building had been built and its ground floor was considerably above the street level grade and the other buildings neighboring it. The reason for the higher building level was the constant flooding subjected to the low-lying areas and living so close to the ocean. I was cynically thinking about how these toilets would look soon in a hypothetical future when Miami is underwater because we haven't done anything to address climate change and sea level rise.

Looking back, I believe I was looking for meaning. Since I no longer felt as if I was working for an international company that would take me back to brazil, I thought I was wasting my time. This American dream stuff turned out to be empty. Around this time my swimming friend Walter had a sudden heart attack during a swimming competition and his untimely death had an impact on me. When I feel death near, I'm reminded of my father's killing. He was gone very early and I didn't spend much time with him. Who knows when my time will come. It could have been while riding my bike to work. When cycling in Miami, I accepted that each time I got on my bike it could have been my last. I didn't want to die before seeing at least a bit more of the world. I wanted my life to mean something. I wanted to do something meaningful.

It was around this time that I decided to quit my job. One day I looked around the open office plan and I couldn't spot anyone whose life I envied. The company boss had just bought a Tesla Model S and was talking about how cool it was. I was happy my stock was going up, but most employees looked miserable and I didn't want to be like them. I drafted my resignation letter and emailed to the CEO and copied Rob, my manager who tried to talk me out of it. I appreciated his concern and I tried to do my best, but I was falling behind, and I was feeling bad about it because I knew I could do better. I wasn't appreciative of my opportunity for which my mother had worked so hard. I could legally work in the U.S., after having lived illegally for two years during high school and here I was squandering all of it. Yet I couldn't do it anymore.

Given my experience hosting Couchsurfers and my construction knowledge, I devised a plan to use green building techniques to build an ecological hostel in Tulúm near Cancún, México. I'd use my savings to buy land and get the first buildings up. From the ant's book, I had developed a serious obsession with insects. After the ants I studied bees and their swarm intelligence provided me with inspiration. The hostel was going to be called Hive Hostel because I felt that travelers acted like pollinators disseminating knowledge, culture and technology. The beds were going to be hexagonally shaped; I even built a prototype in my bedroom in Miami. Some Couchsurfers got to test it. I had to drag two-meter long wood planks and build a single hexagon that took almost the entire room. It was awesome. My mother came one Saturday and took a nap inside of it. I called her the Queen Bee.

Once he received my resignation e-mail, the CEO asked me into his office and wanted to know what I was going to do. I told him I was going to build a Hive Hostel in México. He asked if I was crazy and informed me that construction in México was dangerous and that I was wasting a big opportunity at Plaza. I knew very well construction in México. They had let me work there without health insurance and paid me only one taco per day. But If only he knew how much fun I had had in those two weeks. The CEO was right, but I couldn't stay anymore. Once he realized that I was serious, he told me the company doors would always be open for me to come back. I took that as a good sign, said thanks and left.

I packed my things, my triathlon gear and my bike Cristina (if you ride a bike long enough you will give it a proper name, trust me) and left for Cancún. I donated all the shit I accumulated living in Miami. Packed all my books and put in boxes in my mother's storage unit. I had made the jump. I quit a job paying almost $70k/year to follow a dream. I wasn't passionate about a hostel; it was just the most reasonable explanation I could give to my friends and family. The idea was to build a hostel in a growing city near Cancún. I'd use ecological construction techniques to promote conservation of the environment. The first thing I was going to build would be a compost toilet.

And that's how I went to Cancún on May 5th, 2016. I stayed in Playa del Carmen renting an apartment from a Couchsurfer girl that I coincidentally hosted in Miami. The connections that I could make through that website were incredibly useful. On my first days of freedom, I'd go to the beach in the morning and work on my laptop afterwards. I would go to Tulúm and talk with a real state friend of mine called Abraham and look at possible land plots to buy. I only had 18,000 USD and I started estimating how much it would cost me to buy land and build the first Hives. It was obviously not enough. However, the more I lived there and got to know the market, the more uncomfortable I became with the idea of spending all of my savings there. I was alerted about the insecure legal environment where indigenous tribes were claiming that property sold to foreigners belonged to them and no construction could start in disputed lands.

Furthermore, the legally stable land available was being bought by exorbitant prices to be turned into boutique hotels. Tulúm was a beautiful place and it was attracting luxury tourism, not the eco-backpacker that I was looking to serve at the Hive. Then the veil started falling. There is a famous hotel/bar called Papaya Playa Project that hosts famous DJ's and throws awesome full moon parties. They were desalinating saltwater by using diesel generators and then providing water flushing toilets. Needless to say, that there was none and up until this writing, there is no sewage collection nor treatment in Tulúm. So called Eco-Chique Hotels are free to do what they please with the sewer they generate. If you're on the beach the cheapest legal thing to do is build a septic tank and then pay sewer trucks to pump it out. However, with this being México, who knows where the trucks pump out the sewer when there was no WWTP in the city. There was a lot of greenwashing to appeal to foreigners and I didn't want to participate in it.

Similarly, Bacalar a beautiful lagoon one-hour drive south of Tulúm was already suffering from the raw sewage from hotels and residents discharging freely into it. The Yucatan Peninsula, where Tulúm is situated, is famous for its *Cenotes* which are underground pools created by the local geology. There is no place in the world like it. Yet, increasingly more cenotes were being closed because they were contaminated either by the direct use by tourists or by raw sewage filtering through the porous limestone which makes up the Riviera Maya.

I studied what was happening in Tulúm. Researchers called it "Predatory Tourism." It had happed in Acapulco, Cancún was its latest example and Tulúm was going to be next. Tourists flock to a destination, then governments and business move in to exploit the financial incentives. The city grows, drug traffickers follow and environmental degradation results. This type of tourism was opposite to the sustainable practices that I wanted to promote by building the Hive. We were shitting in the very beaches and cenotes that we came to enjoy.

I needed to find a secure plot of land that I could buy but I decided to take my time. I continued training in Playa del Carmen.

I joined a bike riding team and went riding into the roads cutting the Mayan jungle of Yucatán. I love how sport has kept me in touch with nature. Through this bike ride, I me people working in efforts to make tourism more ecological in the region. I also raced in a triathlon at a fancy beach hotel by cycling the 20 km of a sprint race with my bike Cristina.

I also continued swimming and entered a race from Cancún to Isla Mujeres, a 10 KM (6.2 Mile) open water event. I was joined again by Italian friend Alberto and met a Mexican called Ivan from San Luis Potosí, who later became a good friend. The swim was gorgeous. Imagine swimming for four hours in turquoise water, white sand in the bottom of the ocean, and a bright hot sun. I'm sure I must have swum more than 10km, because the safety boats kept coming out to warn me that I was swimming too far to the right of the swimming line. The line, in this case, is an imaginary line which would be the shortest distance from start to the finish point. Learning to position yourself in the open ocean is an invaluable skill that only comes with practice.

There are no straight lines in the bottom of the ocean like we have in the pool. When we swim, where we look is where we go, and I guess that applies to life as well. During open water events, I must periodically stick my head above the water to sight the buoy or a land reference. When my head is underwater, I cannot look straight down like I can do in the pool. Instead, I must look ahead, into the ocean blue and occasionally scan my surroundings to keep that marine monster or imaginary shark away.

That day, I was slow and ended up swimming alone for most of the time, so it was hard to gauge where to swim. Also, given the curvature of the earth, the reference points on the shore were not visible because there were few short buildings at the Isla Mujeres beach that I could use as a guide. My internal GPS needed calibration. Still, I approached this swim as a practice for Gibraltar.

Around this time, I applied to deliver a TEDx called "The Toilet of the Future" in Cancún which was the result of my research into compost toilets and notably how ancient civilizations including the Mesoamerican Aztecs managed human waste. I became convinced that soon, toilets will have to be waterless for the same reason as in the past: natural resource management. I got accepted and started training and rehearsing for the event which would be in September.

I heard about a Hempcrete and Bamboo construction workshop from Abraham, the Tulúm real estate guy, that would take place in Tepoztlán, near México City and I became interested in it. He and his friend ran a Facebook page called "Heaven grown-Hempcrete México." Hemp is a sister plant of Marijuana, looking similar but without the psychoactive ingredient THC. It's been banned in most countries but it's slowly making a comeback thanks to its practical applications such in construction as this workshop was teaching. I learned that most of the Roman aqueducts were built with concrete made from Hemp or Hempcrete.

Concrete is the most common material used in construction today. It's made from silica through processes which require a lot of energy and environmental degradation. Hemp, on the other hand, can be harvested within ninety days and hempcrete has many benefits such as being lightweight, breathable, a natural pesticide and fire retardant. I started going on a journey to look for ancient wisdoms in practices we had long forgotten. I learned a lot about hempcrete and thought of integrating it into my hostel plans. I enjoyed the course and even accidentally became the poster boy for future workshops because of this picture.

When I arrived in Cuernavaca, the biggest city near Tepoztlán, I went to couchsurfing to find a couch and place to stay. I found the profile of a tall guy called Andy and decided to send him a couch request. He quickly accepted it but when I arrived at Andy's house, I was surprised to find out he was a girl and the guy on the photo was Andy's friend, Jesus. "Okay, cool", I thought. When couchsurfing, you must be open to changes, and that was not a big deal. When I entered the house, I was confronted with a stern-looking motherly figure who introduced herself as *Dra.* Elvira, Andy's mom. I dropped my heavy bags, introduced myself and thanked them for welcoming me. *Dra.* Elvira was apparently surprised and rightfully standoffish. She told me Andy had informed her only hours before my arrival and that I would be their first Couchsurfer. She said in Spanish, "*Esa casa se respeta!*" I smiled and promised I would be their best first surfer. I showed them my profile, references and reassured I would be a good guest.

Dra. Elvira then relaxed and, as a typical Mexican mother, proceeded to ask if I was hungry and had dinner served for us. I went on to explain to them that I had come to Mexico and was researching ecological construction techniques such as the Aztec toilets. Since then, I have visited Andy and *Dra.* Elvira more times than I can count, and I consider her my Mexican adoptive mother. I met the whole family, especially the grandmother or Abu who gave me my blessings. Every time that I am sick while traveling around the globe, I WhatsApp her asking for medication prescriptions. She is actually a gynecologist and it's awkward to disclose that I get my drug prescriptions from my Mexican gynecologist adoptive mother.

From Cuernavaca I went to México City on a one-hour bus ride and every time I was in CDMX, I visited the city center's bookstores looking for relic books about the Aztecs. I love finding antique shops with yellow-paged, worn out and aged books with that unique scent of ancient wisdom printed on paper. When the Spaniards marched on present-day México City, it was Called *Mexico-Tenochtitlán*, and its residents were the *Mexicas*. They are infamously known for their barbaric human sacrifices where during full moons or special dates according to their sun calendar, someone's heart was ripped out. His or her back was placed against a rock pillar, exposing the open chest, which was cut open with an obsidian rock knife, sharp as a razor. The rituals were meant to satiate their gods.

But I got tired of reading about this ritual when I was actually looking for their toilets. I also searched online and couldn't find details, nor drawings. Once I went to the *Templo Mayor*, or Great Temple, which was the major pyramid in the *Zócalo*, main square. This great pyramid was actually the seventh generation with each new emperor building a bigger one on top which appeared as layers of an onion during excavations. When the Spanish demolished the pyramid after the conquest, they used the same stones to build the Catholic church which one can see from the temple ruins.

During the temple tour, I asked the guide about the Aztec toilets and he gave me an answer he was used to repeating thousands of times. He said the *Templo Mayor* also had aqueducts and drains just like the romans. Look how advanced the Aztecs were, they also shit in water, he wanted to imply. However, I pressed further, what about all the other citizens? Did they also have drains? The guide didn't know.

The Aztecs did have a series of double aqueducts leading drinking water into town, like the Romans. However, the Aztecs lived in the bottom of the Valley of México, in the middle of a shallow lake called *Texcoco* prone to periodic flooding. Prior to moving to lake Texcoco, the Mexicas had a vision that they would found their city wherever they saw one eagle perched on a cactus and holding a serpent in its beak, an image that is in the Mexican flag and coat of arms. This dream manifested itself in an island in the middle of Lake Texcoco and that's where *Tenochtitlán* was built.

The Mexicas developed a technique called "Chinampas." You

can still see them in a neighborhood south of México City called *Xochimilco*. The Chinampas were built by driving wood trunks as piles into the lake bottom. They would then fill up these plots with dead vegetation and mud until they were above the water level. These raised beds were used as gardens to produce their own vegetables and their staple food, *Maíz* or corn. Wood boats, today called *Trajineras* were used by the farmers to navigate around the islands and take the produce to Tenochtitlán. In present day, México City, it is customary to take a *Trajinera* tour while visiting *Xochimilco*. The Aztecs literally built their own city lands out of a lake. That's why when the Spaniards first saw it, they referred to *Tenochtitlán* as the floating city.

Unlike, the Romans, however, the Aztecs didn't have drainage, therefore, they could not defecate in the water coming from the aqueducts. The Romans had the nearby Tiber river which conveniently carried the black waters away from Rome's *Cloaca Maxima* sewer system. The *Mexicas* didn't have such an easy method as the lake waters were stagnant, so much that their homes flooded during heavy rains. Furthermore, pipe drains have only been discovered in excavations of *Templo Mayor*. Lastly, Tenochtitlán is believed to have had five hundred thousand inhabitants during the 1500's. Then, what happened to the feces from all of these people, I asked myself.

While studying Aztec architecture, I discovered the common man home had only one room and toilets were shared by a couple of residences or even an entire street. The Aztecs didn't have an alphabet. Similar to the Egyptians they had a system of glyphs or symbols. The glyph representing a house is called *Cali*, which means a box where there is transformation. The society was divided in classes of peasants, farmers, priests and the aristocracy. Only the emperor or the high priests had access to those areas of the pyramid which had drains. Therefore, much like today, only a small elite had their feces carried away by drinking water.

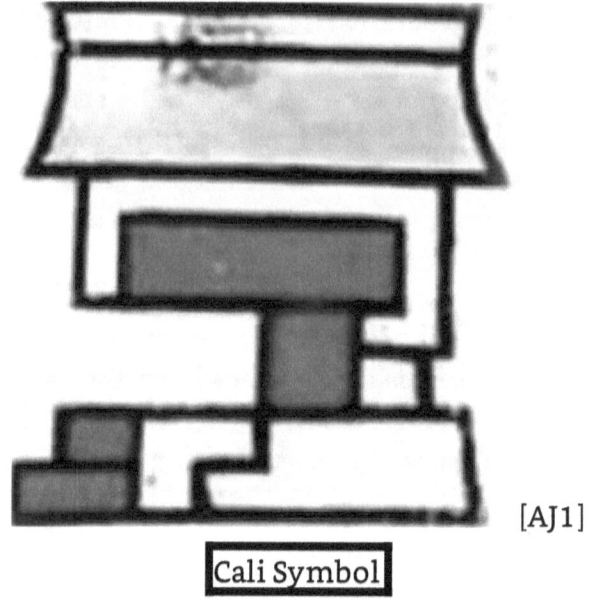

[AJ1]

Cali Symbol

There were no cows in the Americas before the 16th century, they were brought after the Europeans arrived. Previously, animal manure had been the only way to add nutrients to the soil in every corner of the world. In Tenochtitlán, the farmers of Xochimilco used the sludge from the lake bottom to provide nutrients and its water to irrigate the Chinampas. It is said that these floating gardens achieved as many as seven harvests per year and the city was self-sufficient when it came to produce. I did find one footnote reference stating that the Aztecs used human excrements as fertilizers. Human excreta as fertilizer may explain why they achieved such high fertility. However, historians didn't mention whether or not the Aztecs composted the feces, as we know the technique today, before applying it to the chinampas. I'm afraid it's impossible to know.

Once we consider the restrictions imposed by their environment, it is plausible to believe that the Aztecs must have had some treatment of human waste which allowed their civilization to flourish. I surmise that the wastes were collected in pottery jars or straw baskets. Pottery is commonly found in excavations and straw basket would have decomposed centuries ago. If these materials were used, it would explain why there are no relics of "Aztec toilets" per se. These containers would then have been transported by waste merchants to be sold in the main market as some historians have written. Similar markets for human waste have been documented in Asia and Europe under the term "Night Soil." Readily available ash from home fires could have been added to the waste containers to eliminate odors and sanitize the fecal matter. Also, the very farmers could have transported the waste containers from the city blocks on their *Trajineras* boat trips back to the chinampas where they lived and worked.

The Aztecs were known to have high standards of hygiene, bathing daily in the lake and canal system created by the Chinampas. They also had ritualistic sweat lodges or *Temazcal* which I had the opportunity of experiencing a couple of times in my trips to México and even once in Germany. The *Temazcal* domes were built with branches and covered with rugs or carpets. After the bathers went in, prayers would be said and red-hot stones called *Abuelitas*, little grandmas in Spanish, would be brought in the dome center. Then, water was poured on the rocks creating steam; herbs and burnt copal tree resin added fragrances. These *Temazcal* experiences were filled with music rituals and nature symbolisms, in my opinion, reflecting harmony with nature and cleanliness.

When feces are contained and controlled then viruses and pathogens die off in part due to a process called predation as microorganisms fight off each other for resources. Essentially, the worst thing we can do to feces is to place it in water where it can spread. Similarly, the worst thing we can do to drinking water is to defecate into it. Rendering a life-giving liquid into a poison. When it came to sanitation systems, the Aztecs were not able to pollute their lake as they depended on it for their livelihood, fishing and transportation. The Aztecs had to resourceful, not because they fancied sustainability, but instead they were forced to upcycle. They fished in the lake where they lived so water quality was critical to their livelihood. They ate insects and a slime from the lake rich in protein, today known as Spirulina.

All of these factors led me to believe the Aztec civilization can serve us as examples when it came to sanitation. We should imitate their resourcefulness, close contact with nature, respect of water, and nutrient cycling back to the soil. It was sad to notice that most Mexicans didn't know this part of their history, even my scholar friends. I was not surprised to learn that only ten percent of the Mexican capital's sewer gets treated. The untreated 90% remainder flows through *El Gran Canal*, the *Cloaca Máxima* of México, to *Valle de Mesquital* where it is used to irrigate vegetables which fatefully make their way to CDMX's corner taco stands. Lest we continue in ignorance of the Aztecs' sanitations lessons, we risk having to adopt them out necessity. In an uncertain climatic future, similar to the residents of *Tenochtitlán*, we may be forced into eating insect protein, using dry toilets because we ran out of drinking water and raising our street levels because our sea levels continued to rise.

Around this time, I found a website called travel starter which was a fundraising platform to help people start a hostel or travel-related business venture. It was perfect for me. I'd use it to promote the Hive and collect funds to buy a secure land and build the first hexagon pods. However, before doing that I decided to take a little detour to Alaska. Years prior, I had turned down an opportunity of going to study there because I had been offered a shitty job in Miami. To date that had been my only regret in life, so I decided to take advantage of my recently gained freedom and go live there for two months while I got the campaign ready. Alaska had always drawn me for its wild landscapes and incredible nature. I went on workaway which is a website where you meet farmers or homeowners who need a hand and in exchange for helping them, you get a place to sleep and food.

I found a goat farm near Homer and they accepted me. Milking goats is really hard, it turned out. I had milked a cow in my uncle's farm in Brazil, but goat udders are thick, requiring a lot of forearm and finger strength. In the first two weeks I was learning and the remainder two I was crying when I woke up with sore hands and frozen fingers from the low temperatures. By the end, I couldn't even make a fist. There was an Austrian girl working there and she was very strong. She could milk a goat in minutes, pressing hard and squeezing a lot of milk while I looked sad and strained my forearm to get a few drops out.

The family took us on a boat tour to watch whales and out to the river to shoot some guns like we do it in America. I also learned to bail hay which is a lot of work. I followed the tractor spitting out square hail boxes and piled them up in the shed for winter. Farm work is tough. I also helped out in the food garden. From the farm I started the crowdfunding platform for the Hive. I built an explainer video on sustainable tourism and why I wanted to build an eco-hostel I Tulúm and started the campaign. My initial goal was $10,000 dollars which would help me build five hexagon rooms, the kitchen and toilets. However, one month later, I failed to reach the goal and had to go back to the planning phase.

It was almost time to leave Alaska, but I didn't want to go given the fact that I didn't know when I would be back and it's very far. I took another workaway job with a guy who was promising to teach how to build a treehouse. I was allowed to live in a school bus turned camper. It was August but it was already cold, and I had to learn how to run an old wood stove. Having grown up in Brazil, I was not used to these ovens and enjoyed them except when it got cold at four in the morning. I also remember reading Mary Roach's book called "Packing for Mars." Mary is a gifted writer who tackles amazing topics with humor and sarcasm.

This particular book described many of the technologies that astronauts use in space today and what it will take for them to live on the Moon and Mars. I was interested in what scientists had come up in terms of toilets. This was when I learned that the space toilet separates the urine similar to the ecological toilets here on earth. On the other hand, the astronauts' feces and trash generated in space get compacted into a module, think a huge trash can, named *Progress* and it's then shot towards the earth where upon reentry, gets burned to ashes. What I liked the most about space is that it forces us to be conscious about the resources that we use up there and, consequently, here on earth. It's also important to point out the incredible technological advances which were made and then transferred to civilian use. Sadly, sometimes this important benefit gets forgotten and we become wasteful in the name of exploration.

I also got to see the darker side of Alaska. I met seasonal fishermen battling drug addiction and depression. Not everything was as beautiful as it seemed. I went to another workaway job to learn how to build a treehouse. But after a week working on foundation for a real house and wasting time in the rainy woods of Valdez, I realized I wasn't going to learn that at all. I was growing frustrated and the other helper, Eve, who would work as fisherman during the summer gave me some words of advice that I still remember to this day. He said: "You gotta do you man." That was his way of saying, get out of here and go do what you need to do for yourself. With that, I left Valdez by a bush plane, the only way to get around that state. I had to stop by Anchorage where I was hosted by Couchsurfers whose pastime was doing lines of cocaine. It was grim and I got the hell out of there.

Alaska taught me to live in the present and stop regretting past decisions. It was also where I made an important phone call. From the goat farm, I spoke with my Denverite friend Mitch Giraffe and he invited me to Burning Man. I met Giraffe on hangouts a feature from the Couchsurfing mobile app while I was there on a trip to see Anna, Martin's cousin who I met had met in Miami. Giraffe was helping organize a bee-themed camp called Pollen Nation and he had tickets.

I hear it's hard to get tickets for Burning Man, but I wouldn't know because I didn't try to get them; they came to me and I took the chance. I've learned to live like that since I started traveling. I asked Mitch if I could also bring my buddy Rei to which he said sure. We gladly paid the ticket and camp fee which were around $650 USD for each of us. The festival might as well be called burning money, but we knew it would be worth it.

I met up with Giraffe and Scott in Denver, and I helped them with last minute preparations. Giraffe asked me if I wanted to come on the build team and help set up the camp. I said yes I would be able to go earlier and avoid the infamously long wait lines. We went to a military surplus store to buy stuff we'd need for the camp. Burning man takes place in a flat desert that's called Playa made up of a fine grain called Playa Dust that will get everywhere, I mean it. It's advised to bring snowboarding goggles, and facemasks to protect oneself. I bought a bandana and got a mask at a second-hand store. Then we bought food at Walmart near the burn, as people who have been to the burn too many times, like to call it. Since this was my first time, I was called a virgin and if you come unprepared, then you're called a *sparkle pony*. I didn't want to be a sparkle pony. There was a lot to learn. The festival is one week-long, but the first days are slow since the camps are still being built. We drove from Denver to Nevada, going through Utah. It was an awesome road trip where I got to hear lot of stories from the burn, but I still kept my mind open to what I was going to experience there.

Another benefit of coming on the build team was to avoid the infamous long lines getting in. I heard about people waiting up to ten hours in line. The good thing is that the atmosphere was so relaxed and even before getting dusty I could feel this was going to be an awesome festival. I had never had a real festival experience and burning man was going to be the first one. I did go to a Christian festival near Orlando called Cornerstone where I listened to Christian punk rock, but it was a very clean festival; we stayed in a hotel. There were no hotels at burning man, the closest city would be unreachable for ten days.

We got to our camp at 7:15 and H. the city is shaped in concentric circles going from the inner one at A and going as far out as J. The numbers are from the dial of the clock. We started building domes, setting up the camp tents, kitchen and bar. The first person I met was Lisa. She had arrived a couple days earlier when there was no one nor any shade structure and was already suffering from second degree burns. Poor Lisa. I met a lot of people in the next days and learned about a lot of news things in a very small amount of time; it was overwhelming. One of the core principles of the festival is radical self-expression so there were people wearing all kinds of clothes or not.

Another aspect of the burn is the incessant amount of jokes and double entendres. Scott brought a game called "what's in my Butt" where there was someone's butt cut out with a hole and black sleeve in the anus so that one could reach in a secret box and pick one item that someone else left in there previously. There were a lot of jokes about this while I was there. In one of the cross streets, there was a pole with a "Flesh-light" and a sign above that read "Public Flesh-light." There was a kissing booth with no one in it which meant it was open to anyone who wanted to sit there. There was a human carwash where you could go and get the dust washed up by strangers. There was a famous orgy dome which I never saw. There was a kissing roulette game that I went in the hopes of kissing someone and ended being told to kiss a guy for the first time. I should have been more specific with my wish and got back in the queue. The second time around, I was told kiss the first girl to my right who turned out to be a large lady with piercings in her mouth and nose. That was also a first. This was only a Tuesday.

We also got placed right next to the toilet block. There were about twenty porta-potties diagonal to our camp. By Wednesday, I didn't need to see them to know where they were, as long as the wind didn't blow in our direction. The cool thing about the burn was that everything was allowed as long as you had consent, that is permission from whoever you were interacting with. Otherwise you were free to do whatever you wanted by yourself. I learned to ride a bike, naked. I listened to great music sets.

There was a fire conclave in our camp. They were training all the time for their respective fire shows. Meanwhile, we were building the camp and getting to know each other. Drama and love ensued. There were also artists running around trying to set up their art pieces or projects, people looking for tools. It was chaotic and beautiful. I met so many amazing individuals on different journeys. I was starting to learn why no one could really define this festival. One night early in the week we were walking by a neighboring camp and there was a woman wearing black leather clothes and spanking people passing by. I gladly volunteered and Lisa started calling me "Spanky" after that. She also said that would be my Playa name for life.

My best friend Rei arrived. We had met in college, had been flat mates in Miami and he came to the pool that day that I was panicking. However, he soon got lost on his own burn. It was so hard to meet friends or make plans to go to an event because there was so much going on. Luckily there was very little phone reception at that time and that is one place you don't want to be on your phone. We would run into each other and just amuse ourselves by recounting what we had seen, heard or experienced previously. There were "mutant cars" which are vehicles that were approved to roam around the burn at slow speeds. These are also called art cars and they have different kinds of designs, some spit fire.

Once I saw a car that was the little purple monster character from Pac-Man. You know, the one that eats Pac-Man? At night, the lights and the playa dust made everything look otherworldly. One night, Rei and I were talking by the fire in our camp and one magic-carpet car drove by. It was just a flat surface with a huge carpet and the wheels hidden beneath it. When Rei spotted it, he shouted: "look at the magic carpet!" ran and jumped on it. I didn't see him for a few days after that. I also went on the magic carpet one night and talked to a girl named Jess who worked for Clifbar. I told her about my key west swim and, before jumping of the flying carpet, out we kissed.

The thing about the playa is that it is flat which makes it great for cycling. I've always loved cycling. I can still remember the first time I learned how to ride. my mother taught me in the back streets of Sobradinho. The houses in my city had two entrances, the front entrance which led to the paved street and the back entrance led to the "faixa verde" or green zone. This zone was like a communal backyard which all the houses on the street shared. There was no asphalt, only sidewalks and in the middle grass and a dirt road. My mother taught me how to ride by pushing me and running next to the bike. I treasure the memory of being able to balance myself on the bike without the training wheels. At first, I didn't know how to brake so I just coasted off the sidewalk and fell sideways onto the grass.

At the burn, although it had no grass, it was also safe to fall. Luckily, I didn't but I had to remind myself all the time that *"the rules of physics still applied."* The atmosphere was incredible, the heightened states of mind thanks to all the art, music and drugs was contagious. It felt as if there was no limit to fun, freedom and self-expression. That's also where I rehearsed my TEDx, somewhere in the playa, standing in front of a two-meter-tall speaker, feeling the soundwaves traveling through my body and tripping on a concoction of chemicals. That was just a bit of burning man. It cannot be defined; it must be experienced and I'm glad I did. I had spent ten days there in total and by the end I was getting a bit crazy. I needed to get out of there running, semi-naked. To make matters worse, the toilets nearby were getting filled up and the cleaning scheduled couldn't keep up with the booming population. Each day thousands more residents arrived. Some estimate the burn to have as much as 70,000 people. On the last days there were dust storms where I couldn't see 2 meters in front of me and that was the feeling that I got out of the burn, a shitty dust-storm.

I got back to Denver with a new friend Laura and hang out there for a few days before heading back to México. On the way, I learned through Chris, a French American space engineer who I had met at a couchsurfing event in Playa del Carmen, about the Aerospace International Congress that would be taking place in Guadalajara, México. Coincidentally, Efra, my Peruvian polyglot friend who introduced me to couchsurfing was also in Guadalajara. He had moved there after getting accepted into an accelerator program with his start up and I stayed in his couch during the week that the conference took place. I don't remember how much the conference cost, at least one hundred dollars which doesn't seem like much but at the time I was already concerned about my savings and I made every effort to save money, so I decided to crash the conference. Given my positive track record crashing events in México, I carried my F1 badge with me at all times.

After México, I realized I was getting a huge adrenaline rush from doing this, so I went to the conference without a ticket. When I walked into the center, I spotted the registration booth on the right and walked confidently past the security guys and volunteer girls who were checking badges. The key to crashing events is to look and act like you belong there; talking on the phone and looking busy also helps. Once or twice a security guy may have asked me about credentials, but I was too on busy the phone to notice him in the corner of my eye chasing me down, his tie flapping over his shoulder. !*Señor, Señor!*

I remember first being interested in space exploration when I watched the Space Shuttle Columbia disintegrating while re-entering the earth's atmosphere in 2003 on TV. That's when the vehicle experiences temperatures of 1,650 C (3,000F). Coincidentally, or not, I later won the Ronald E. MacNair fellowship, a research grant named in honor of one the deceased astronauts in that crash. Unfortunately, I didn't conduct any research through the fellowship. I may have been too late in the application process and I didn't find any research proposal or lab program from the National Science Foundation that interested me.

If only I was into toilets back then.

I thought it was worth going to that conference because the world's space agencies were in attendance and I wanted to know if there had been any advances in the space toilet system that could be applicable to earth or were relevant to my upcoming TEDx. Also, I wanted to go because I heard Elon Musk would be there to reveal his plans for Mars colonization. I even cheekily posted on Facebook that I would talk to him about toilets in Mars even though I had no idea how I would be able to do that. I'm not afraid of talking to celebrities or people in power. Once I got an autobiography signed by Brazil's former president, Fernando Henrique Cardoso in an event in Atlanta, Georgia. He had been president while I lived in Brazil and at the time, I never imagined I could be in the same room talking to this seemingly important person. It was as if in the United States, everybody was equal. I thought I could also talk to Elon Musk about his plans for Mars colonization and why not ask him about toilets in Mars?

When we land on Mars, we will need to shit as well. Once I read that when the astronauts left the moon, they had to eject their diapers on the lunar surface. Therefore, if you go to the moon right now, in addition to the flags, the moon landing modules, you will also find old astronaut shit in adult diapers. They had to be left there because it's important to shed any extra weight before the rocket can take off and that's why shit stayed in the moon. Is that what we are going to do in Mars?

During the first days of the week, I visited all the booths and talked to all the space programs and realized that shit was not a high priority. I got to see the diaper that astronauts use on the way to the ISS and leave on the moon. However, for longer space explorations such as going to Mars, it won't be possible to shoot shit towards the earth because we will be too far. Unless they hope on shooting Shit-Filled-Progress-Modules into the void of space all the way to Mars? The journey there will take at least six months. I was hoping that the space agencies were thinking about composting the feces on the way there and using them to grow food once the astronauts actually get to Mars. That would be the easiest thing to do, I assumed. Haven't they watched *The Martian* film, I asked myself.

Thanks to SpaceX, it costs now about $3,000 USD to send one kilogram to space, but that number used to be a lot higher like $54,000 back in the early 2010's. It takes a lot of fuel to escape gravity's hold and get to space. This is why the astronauts are forced to treat their own urine and drink it up in space. I thought that if they can upcycle urine, then why couldn't they also compost feces? I asked that question to the Indian space agency, the European, and all the other booths that would listen to me. Still, I didn't get any real answers. Except for the Japanese agency. Perhaps since they live on a tiny island, they were more used to being conscious about their resources. I also talked to Orion staff which is the planned capsule that will take American astronauts to the ISS. Currently, since the U.S. shuttle program was dismantled, only Russian rockets can take astronauts to the ISS. This reliance on Russia is one of the reasons that, in my opinion, NASA was eager to support Elon's SpaceX, so a U.S. company could take astronauts to space. SpaceX is poised to beat the Orion program in actually taking the astronauts to the ISS.

I didn't learn anything new though and I grew frustrated. I looked toward Friday when Elon was going to speak about the plans for Mars. After he talked, they opened the session for questions, and I took the opportunity to ask him about the toilets in Mars. I was preparing the question by writing on a piece of paper and it was nerve-wrecking. I made an embarrassingly long question that you can listen for yourself on YouTube under the heading "Elon Musk gets Bizarre Questions." It has gotten one million views. I thought I was doing standup. In essence, I related Mars to my Burning Man experience because it had been dusty and shitty and he's been to the burn, I read. But I seriously wanted to know if sanitation was a concern for those thinking about colonization of another planet and if we could use the mindset of not polluting Mars and applying it to our home planet. I made a lot of assumptions.

Elon looked a bit surprised at first but then said there is frozen water in the Martian poles. When he was answering and minutes afterwards, I remembered feeling a bit lost like someone had pulled the floor below me or shell shocked. His answer showed that he wasn't applying the very "First Principles Thinking" that his biographer wrote about. To recover ice from the Martian poles, melt, then pipe it to a station so that astronauts can shit in it and then treat the blackwater takes a lot of energy that will be valuable in Mars. I seriously doubt that if we get there, we will be using water flushing toilets. Even NASA simulations of the Mars mission in Hawaii have used compost toilets. Up until this point, I confess I wasn't very clear about my goal in life or with toilets but that day something clicked. The world's leading scientists and Elon Musk didn't know shit.

Unless of course, we used the same system as some of the research stations in Antarctica which are dumping raw sewage in holes dug in the ice or pumping it out in the ocean. I mentioned some of these systems in the TEDx and I confess I was shocked to learn that the same scientists warning us about global warming are also polluting the waters in the most pristine continent of the planet. How can we "save the planet" if we can't even save our shit from polluting our planet? Couldn't we save energy and water by using compost toilets? Why do we always assume that it's up for someone else to change their behaviors while we continue spoiling the only planet we know which supports life? Nothing made sense to me anymore.

I went back to Cancún and delivered the TEDx on the toilet of the future at the end of September. My long-time friend Diogo flew in from Miami to see me deliver the talk. It was a great experience and I'm thankful to the TEDx staff for all the training and time invested into us. I wasn't very nervous before the presentation. Thanks in part to the shot of Mezcal I had prior to going up on stage. I remember once walking outside of the red dot onstage and that's when I lost my trend of thought, so I went back in and resumed the talk. I think they edited it out. I'm not sure, I haven't been able to watch that talk on YouTube since it's been posted. I'm too embarrassed to see it. I'm thankful to the open mics and improv theater sessions that I've done for my ability to go up on stage and talk to strangers. It's an invaluable skill that I will keep on developing. I believe that my urge to transmit my ideas is bigger than any self-consciousness I may have. After the event, Diogo and I went to visit the Mayan pyramid of Xichen-Itza and I flew back to Brazil to spend time with my family which I had not seen in a while.

5- SWIMMING
IN OAXACA

After reading Joe Jenkins handbook, I became more interested in Ecological Sanitation and started reading other books from its bibliography. One name kept coming up, it was Carol Steinfeld author of books on compost toilets including a notable one called "Liquid Gold," about how to use urine as a fertilizer. After reading her book, I found her on Facebook. I remember asking her questions which she promptly replied, and we've kept in touch since then. We've even met in person during a quick trip I made to San Francisco nearby where she lives. She has become a mentor to me and I'm thankful for her patience and guidance over the years.

Another person whose name kept popping up was César Añorve, a Mexican architect who lives near Cuernavaca where Dr. Elvira lives. The first time I went to Cuernavaca to attend the Hempcrete workshop, I took the opportunity to also visit Mr. Añorve's house. He has been building and championing composting toilets since before I was born. Once I witnessed him giving a workshop to young students where he referred to the porcelain toilet as the *Ángel Exterminador* or "Angel of Death" and I must confess I found the statement rather odd and dismissed it as coming from someone who's been in the shit business for too long. However, since then, I have come to agree with him.

A couple of houses down from his showroom, there was a small workshop where his workers manufacture fiber glass urine diverting pans and toilets. They look white like the common porcelain toilet but instead of connecting water pipes, a plastic bucket is placed inside the toilet to collect the feces. A small water bottle is placed under the urine diversion section where urine is diverted with help of a small hose. Once one uses the toilet, then he or she is instructed to cover up the feces with a mix of sawdust and ash. This ash/sawdust mix is used to prevent flies from landing on the excrement, controlling odors, soaking up the water excess liquids thus controlling the moisture content and, finally, to balance out the nitrogen and carbon.

I learned from the Cornell University Composting guide that in order for composting to take place, we must reach a proper Carbon/Nitrogen ratio. Feces and urine are rich in nitrogen and the sawdust is rich in carbon. Once the bacteria in the excrement start decomposing the feces, heat, water, carbon dioxide and humus is generated. Essentially, as the bacteria from feces are breaking down or decomposing organic matter, they heat up the compost pile and this heat kills pathogens and unwanted microorganisms.

Also, from the Cornell Composting website, I've learnt that there are three phases to the composting process: 1. Mesophilic Composting or moderate temperature, 2. Thermophilic Composting or high temperature, and 3. Maturing and cooling phase. During the mesophilic phase, the temperature in the pile reaches 45ºC and during the thermophilic up to 70ºC. Temperatures this high for prolonged periods of time will sanitize the compost material killing viruses, roundworm eggs and pathogens. In fact, as per the Environmental Protection Agency, EPA, guidelines five days at 40ºC and at least four hours of them at temperatures above 55ºC will ensure that the compost generated is safe to be used.

Nature has its own way of up-cycling and reusing the by-products of digestion. It's up to us to incorporate composting science and technology into our live styles in order to live sustainable lives in harmony with our environment. Otherwise we risk not only polluting the planet but our own livelihood. I believe that the consequences of global warming and climate change are the direct result of our inability to live according to the laws of nature. In Joe Jenkin's Humanure, he draws a parallel between the common fever and global warming. He proposes that we, homo sapiens, are acting like an infection to the planet and much like our body heats up in fever to kill viruses and infections so is the planet heating up to get rid of us. Considering how many people have been and are slated to be killed because of global temperature rise and its consequences, this was a powerfully simple argument with which I agreed.

I went to an event called EcoFest held in Mexico City, a gathering environmentally conscious companies, NGO's and speakers. It was there that I met workers from Heifer-Mexico and found out about their humanitarian projects in Oaxaca. When you're in Mexico, you will surely hear about Oaxaca. It is a special place with a rich culture, cuisine, and an authentic soul; it is also one of the poorest states in the country. They explained to me that they were helping families by giving them twelve chicks and at the end of one year, the receiving families were to donate twelve chicks to their neighbors and thus continue helping one another. Heifer also supplemented children's diets in case of malnutrition and made sure they were going to school by visiting the communities weekly. I liked their approach and told them of my wish to donate a dry toilet to a family of their choice, if they also included keeping an eye and making sure the family would be using it. I knew that many sanitation projects have failed because after receiving toilets, the beneficiaries stopped using them, since they saw no real benefit in defecating in a toilet when they could just go outside. Heifer Mexico understood very well the issue and agreed to follow up with the family after the toilet had been donated.

I flew to Oaxaca for a one-month stay. I found a "Bee-Lady" on Couchsurfing and ended up renting her spare room for two weeks. She was a Swiss expat and had been living in Mexico for twelve years. She came down to write a thesis on the native Melipona stingless bee and stayed. Her boyfriend was a Mezcalero master who taught workshops on how to make sustainable Mezcal, a distilled spirit. She had three dogs she adopted from the streets and occasionally sacrificed some honeybees to practice apitherapy. I received the treatment twice. She would grab a buzzing bee from a glass jar with tweezers and carefully place the doomed bee on my knee which it would sting. The pain was bearable to me but sadly, deadly for the bee. Every time a honeybee stings someone, it dies because it cannot pull back its stinger, instead its digestive tract, nerves and parts of the abdomen stay connected to the stinger.

There was a gym nearby, where I got a one-month membership, so I could train at the 25-meter pool. I still had no idea when I'd be able to swim across the Strait, but every time I entered that pool, I told myself: "This practice is for Gibraltar". The water was colder than usual, and I told myself to get ready for the fresh waters of the strait. I also started working out with weights and cardio machines to keep in shape for any impromptu triathlon. An added benefit of staying fit while traveling is being able to chase down any bus while carrying my backpack.

Long distance swimming is an endurance sport. That is, one must be able to keep on swimming for a long time. In order to train my endurance, I looked to the northern mountains surrounding Oaxaca for help. I learned of many ultra-marathons that take place in those hills and I looked to train with a local runner. I found a Couchsurfer who ran long distances, his name was Fernando and I sent him a message on the website and asked if I could accompany him on his next long run. He promptly agreed and we set out on a Saturday out of Tlalixtac de Cabrera, the small village where I was living with the Bee-lady and her three dogs.

We started walking and when were warmed up, started running up the *Cerro* or mountain in Spanish. He told me of his last ultra-run which lasted more than twenty-four hours where he went hungry and after the sun set, started hallucinating with elves or *Duendes del Bosque*. Maybe they were real, he confided. I was glad we were only running 37kilometers and during daylight. Some parts of the hike took us alongside a creek, and we met some mezcaleros carrying freshly cut agave plants with help of their faithful donkeys. After a couple of hours running up the hill, we had a well-deserved break at a small restaurant where we had *Tlayudas*. The food in Oaxaca is amazing and *Tlayudas* are something everyone will try while there. It's a large thin and crispy tortilla with refried beans, herbs, guaca-mole, meats and *Queso-Oaxaqueño*. They are very proud of their cheese down there and it's not without merit. After our banquet, I rested my legs by propping them up a pillar outside the home restaurant patio getting them ready for the downward journey back home. I hoped this nine-hour hike would serve me as both mental and physical endurance training for my swim, even though I had no idea when it would take place.

One week later, the Heifer staff took me to a village called Santa Ana del Río which is a two-hour drive outside of Oaxaca city reached only by a gravel road, snaking around precipices and canyons dotted by cacti. The agave plantations along the steep mountainsides seemed as if they were drawn by someone ignoring the laws of gravity or with no care for the poor souls who would have to tend the diagonally planted crops.

Once we arrived in Santa Ana, I met some of the Mezcal producers and families being assisted by Heifer. Some of them were drunk even though it was only ten in the morning. Mezcal is the predecessor to Tequila. In Oaxaca, it is made by taking the agave plant, cutting the spades leaving only the *Piña* core, which is then cooked over 24-hours inside a three-meter deep hole in the ground, covered by blazing volcanic rock just like the *Barbacoa de Borrego*. The cooked *Piñas* are then transported to a circle track where a donkey pulls an ancient rock wheel crushing and squeezing the juice which will be fermented, doubly distilled and left to age in wooden barrels. The result is Mezcal, a powerful spirit with earthy and charred tones which will make you fall in love with Oaxaca to which the bee-lady could attest.

We walked up a dirt road alongside another mezcal-producing shed, past a resting donkey where I met the family that was getting the toilet. They had two children, a boy, and a girl, and the four were getting ready to move into their new home, still under construction. It was a single building with a large room, no divisions and the red bricks were exposed like my house in Brazil. They didn't have a toilet room or shower basin, so we brought the urine diverting toilet seat I had purchased in Cuernavaca from Mr. Añorve which had been installed onto the wooden box which I had made by a carpenter in Oaxaca City. We placed the compost toilet in the middle of the living room so I could give them instructions on how to use the toilet, in Spanish.

whereisaldoj
Santa Ana Del Rio, Oaxaca, Mexico

whereisaldoj Santa Ana del Rio in Oaxaca, México. Typical family in the valley who survives on subsistence farming and the local #mezcal. #whereisaldo #Travel #ikigai #digitalnomad #México

little.magics You're god or Jesus or something

explorophile Wow! Where is this?

Aimé par oscar_srna22 et
74 autres personnes
18 AVRIL 2017

Ajouter un commentaire

I asked them where they would like to build a compost pile with the bucket's contents, but they opted instead to just bury the shit in their backyard. Technically, that would not be a composting pile, but it would be better than flushing raw feces into the stream running next to their house. I also accepted this as a small improvement and once they were familiar with using the toilet, then we could teach them how to manage a compost pile. It was important to me that they would be comfortable with the toilet, so that they would use it. There was Heifer staff in the village familiar with dry toilets and they assured me they would be following up with the family. We took some pictures and then went to eat home-made delicacies and grabbed a bottle of the best Mezcal I'd tried until then. On the sinuous ride back, I was feeling nauseous. I wondered if one toilet could make a difference. I reminded myself of the most famous statistic on sanitation at the time, "2.3 billion people don't have access to sanitation". I knew this figure was more than ten years old and I guesstimated it must have surpassed three billion.

I once read a book about *Terra Preta do Índio*, or Indigenous People's Black Earth in Portuguese. Scientists had recently discovered that where there were signs of large civilizations in the amazon, the soil was dark in contrast with the red, rich in iron, more common soils in the amazon basin. After further excavations, they realized these civilizations deliberately made the earth black my mixing charred organic materials. The scientists called it Biochar or biological charcoal.

Growing up in Brazil, I visited my politician uncle's farm in the interior of the neighboring state Goiás during my school breaks. I loved that farm so much that I hoped I would one day be a cowboy, riding horses and herding cattle into the sunsets. But I don't think my mother and aunt liked my plans because when I told them of my ambitions they suggested me to become a veterinarian. During these trips, I'd see huge termite nests which got as tall as 1.5 meters, or five feet. These structures called, *cupinzeiros*, looked like giant red mounds in the middle of the pasture. Additionally, there were other red earth structures similar to these cupinzeiros, although larger and man-made. They were called *carvoeiras*, or charcoal makers, and were built like red igloos, dome shaped, with red earth bricks by local men who lived in the farmlands. These farm hands were simple, hardworking guys who most likely never went to school, but they did teach me how to make biological charcoal.

They would go into the forest, cut down thin trees and let them dry before chopping the branches into small segments that could be fed into the charcoal makers. Once the *carvoeira* was filled, they would cover up the openings, leaving only a couple of breathing holes throughout the dome. They would then light up a piece of paper on fire at the base opening and seal the dome which would smoke for a few days. I was learning how to make bio-char.

The difference between making ash and charcoal is a process called pyrolysis. The wood was not being completely burnt or combusted. Instead, the dome had few openings to precisely limit the amount of oxygen available to the fire. Charcoal was being made by heating up the wood or organic matter in the presence of little oxygen. The heat was evaporating the water and cellulose from the wood and the black charred wood remained. As the water molecules evaporated, air pockets were created, increasing the surface area of the newly made charcoal. Think of it as a hotel with lots of rooms. Its large surface area per kilogram of material is what makes charcoal an excellent filtering material. Scientists also discovered that when mixed with the earth, the charcoal retained water and healthy soil bacteria in the pores, or rooms from the hotel metaphor. Plants grown in *Terra-Preta*, black earth, could access water, nutrients and healthy soil bacteria on demand. This black earth enabled the creation of large human civilizations in the Amazon Basin, something not thought possible until recently.

Biochar from human feces could be used as one of the solutions to the global shit crisis but it's important to note that this is a high energy demand option and that carbon dioxide, a greenhouse gas, is produced during the production of biochar. At the time of this writing, I'm calculating the Energy-Mass Balance of this option so that I can determine whether making biochar from feces at large scale events such as Burning Man and other festivals around the world is truly a sustainable alternative.

After having traveled to thirty countries and gone to music festivals, I've unconsciously equaled the two. For example, when you go to burning man you must be self-reliant; you have to bring your water, food and shelter and when I went to India, I had to be careful about all of them. The only thing I could expect "for free" from the festival organizers was the stinky porta-potties. Obviously, they were not free, the cost was covered by the ticket which cost around four hundred dollars. When I travel, I like to think about governments as festival organizers.

India, for example, is a chaotic festival to me. The decoration is unparalleled, but the water could be safer.

When governments or festivals are mismanaged, the citizens and festivalgoers must take responsibility for their own safety and well-being. Can you imagine if your country didn't provide potable-grade water for you to defecate into? Stop and think about that for a little. If you, living in a rich country of Europe or North America didn't have clean drinking water at the turn of a tap, would you have a water flushing toilet? I'm sure you wouldn't. It would be far easier and cheaper to have a dry bucket toilet and to take care of your shit in your backyard or patio.

I would like to demystify one aspect of traveling to developing countries in case you're not familiar. When I grew up in Brazil, I had access to water flushing toilets. When I go to México City, Dakar in Senegal and Addis Ababa in Ethiopia, I also have access to these "European toilets." All the hostels which I've been in India also had water wasting toilets in order to cater to well-paying westerners. Toilets are available but only to us, the minority of the world's population. Every time we use these toilets, we are perpetuating inequality between the few who can afford to pollute drinking water and the majority of people who don't have safe water to drink. If there is anyone who must use compost toilets now, it is you and I because we have access to the technology, the funds and the luxury to choose what happens to our wastes. The few families to whom I had the opportunity to donate a compost toilet were too busy worrying about having a job, buying medicine or putting food on their table. When I showed at their doorstep to give them a compost toilet, they welcomed me with a type of skeptical gratitude.

This questioning led me to conclude that governments while attempting to improve the quality of life of its citizens enable wasteful behavior. Once I realized this, I started to appreciate "bad governments" or "third world countries." I have actually completely changed my opinion on development, and what being a rich country means. I've had a hard time building compost toilets in "poor countries" because its citizens want a water-flushing toilet just as much as rich countries' citizens. A poor Mexican doesn't want to take care of his shit just as much as a rich North American who enjoys flushing and forgetting his daily bowel movements.

Each time that you brew coffee or use an electric toaster for breakfast, you are using more electricity than people in poor countries will use in their whole day. I've come to realize how hypocritical it is to force technologies such as compost toilets on people who are struggling to survive while we are giving them the worst example. Our movies, marketing and propaganda show the entire world that in order to be civilized, they must cut down forests, consume a lot of electricity and wastewater.

In the west, we've created government so that it will, among other things, take care of the dirty work for us. The government is supposed to take care of the poor, wage wars to guarantee our precious resources and take care of our wastes. Countries that don't provide this luxury to its citizens are therefore called "poor countries" or "developing nations." Yet, the more I travel, the less I want Brazil to be like the United States nor India or México. What makes the world a beautiful place is the diversity within it. I believe that less government and more individual freedom is what will ultimately guarantee an improvement to standards of life.

Education is a good example. Thanks to the internet and mobile phones, we have access to unlimited education. Take how I learned French with Duolingo. However, when I travel to poor countries, it saddens me how many restrictions governments impose and protect their markets from international mobile service providers which would increase competition and lower prices. India is an example of how more competition between phone companies lead to widespread access to mobile data. Ethiopia on the other hand, shows how monopolies protected by the local governments restrict access to information and keep the population uneducated.

Going back to water, it's well known that in mid-19th century Europe citizens of London and Paris would throw buckets of shit on the streets and when water was made available in the household and toilet bowls became possible, then the cheapest solution was to channel sewers to the local rivers. What if the governments had not allowed sewer to be disposed untreated in the rivers back then? What if you had to treat your own sewer? Would you still have a white porcelain toilet?

6-SWIMMING IN ENGLAND

In the summer of 2017, compost toilets took me back to Europe. While researching toilet companies, I became familiar with and reached out to Hamish from Natural Event. He's an Australian guy who started the biggest compost toilet company in the world. He provides compost toilets to the largest music festivals in England. I contacted him on Facebook messenger after having watched his TEDx on shit titled "Giving Back Shit its Good Name." In his talk, he speaks with his distinct Australian accent about the compost *Revolootion* and its power to transform how the world poops. I thought I had met my shit hero. He ends the conversation by saying "next time you take a poop, think about me." What a man!

I asked to work with him that summer and he asked me if I smoked weed. I reassured him that I didn't. He said it was okay if I did as long as it wasn't during work. I reassured him that it wouldn't happen, and he agreed to take me in. I had actually tried weed in Miami before, but I never got high or was drawn to the drug. Also, working in construction I was subjected to random drug tests. There's a history of drug addiction in my family and I didn't want to become another addict, so I had stayed away from all drugs except alcohol. I packed some clean socks as per Hamish's instructions, bought a sleeping bag and brought those heavy and yellow FE prep books which I used to study for the exam the first time. I figure I'd have to eventually re-take the exam and get my diploma, even if only to make my mother proud, especially now that I was going to be cleaning toilets in Europe. When I realized that I was going to be near Gibraltar, I sent Rafael an email asking about swim placements, but I got no reply.

I flew into London and took a train out to the closest city near Glastonbury, the season's first and biggest festival where Hamish picked me up at the station. My new co-workers were interesting people. There was one crazy Italian guy who liked to smoke a lot of weed. There were two Hungarians who became best friends. I really enjoyed working with them and British people; the accents and sarcasm made shit work hilarious. They were a band of misfits. Most traveled around the world during European winter and made money to pay for such trips with shit at festivals during the summer. It didn't take long before I relaxed myself and got to know them. During my first week, I even tried studying for the FE at night after my shift, but that soon revealed to be very naïve of me.

First, I had to learn how to assemble heavy steel and alumi-num frames which were stacked in flatpacks for transportation. These sets required at least three people to set up a toilet. The first three weeks were devoted to assembling one thousand toi-lets at Glastonbury, the summer's biggest gig. I was either on the bowser team, using a pressure washer to clean the toilets or on the building squad. The toilet was simple to build. Single frames connected to each other to form blocks, ranging from two to twenty toilets. The roof and walls were made of heavy-duty plastic similar to circus tents. Underneath the toilets there were 200-liter trash bins connected in series with garden hoses. A handful of hay was placed at the bottom of each bin to filter out the urine, keeping the shit from clogging the urine hoses. At the end of each block there would be a bin with one sump pump sucking out all the "*shiss*" from the block into a

1000 liters IBC, Intermediate Bulk Container. Shiss was the precise term my new colleagues chose to refer to the "urine filtered through shit" or as sanitation experts call it "leachate." Most of the waste in festivals is urine. A good toilet system must be able to separate the two in order to last longer with minimal servicing. I have yet to see a better urine diverting system that can withstand the abuse and misuse by festival goers at their different states of mind than the one at Natural Event.

In the next three months working at music festivals, I saw, handled, shoveled, and smelled more shit than most humans will produce in their lifetime. It was tough and incredibly disgusting but, at the same time, it was also a priceless experience. I learned a lot about human behavior, the logistics of shit handling at large events and working on the move. Also, besides the tonnage of human feces, I had never seen so many types of drugs in my life. This was going to be my second festival, with the first one being the burn.

Many users tended to drop their drug stashes while in the toilet and every time we cleaned them, there was a chance of finding something. There was a gap between the floor and the seat panel where coins and bills also fell through. Going to clean the toilets was always a surprise because you never knew what you'd find. After three months I could tell most drugs apart just by looking at them and I had finally learned how to properly smoke weed with my shit hero, off the clock, obviously.

When Glastonbury started, we were given leadership over cleaning teams where crew like me, were responsible for teams comprised of volunteer cleaners. These festival goers got free tickets in exchange for twenty hours of cleaning during the festival. Glastonbury tickets were expensive, hundreds of British pounds and hard to get and being on the clean team was an "easy" way to get a ticket. The hard part was herding the volunteers which partied too much and didn't show up to work sober or at all.

But I was happy working in the "Clean team." I couldn't believe it, but cleaning toilets made me happy. At first, it felt weird walking up to the toilet blocks with rubber gloves on, cleaning supplies in a bucket and cutting in front of all the people in line to clean a toilet. The festival goers would just look at us with awe and respect. They were happy to wait to be the first one to use the toilet after I had cleaned it. They would often thank us for doing this job and I could tell from their facial expression that they would never do it themselves. It was as if they thought that we were brave for cleaning toilets. Somehow that gave me pride. Yeah, I'm brave, I thought. I'm not afraid of shit and I wasn't embarrassed either.

My mother also cleaned toilets in the U.S., that's how she paid for my private education in Brazil. There was no reason to be embarrassed about it although when I was young, I must admit that I may have been. It's a silly feeling to be embarrassed about cleaning toilets. It's an honest work. Yet, somehow, I was taught to feel that cleaning shit should be a job left to those without education or people who made too many bad decisions in life. Sometimes I would wonder if had done the right thing. I had at least one diploma from a U.S. university I'd reassure myself, and I was very close to getting a second one in engineering and here I was cleaning fucking toilets in England. What had I done with my life. I was making good money in Miami I reminded myself, yet I was having so much fun with these people I had just met. But I knew that this experience was enriching my life in more ways than I could realize at the time. I trusted my path and kept cleaning.

There was one volunteer in particular who really impacted me. Her name was Chaz. She had blonde dreads, piercings, tattoos, and a daughter. She was great to work with. She was so humble, such a hard worker and also knew how to party. Her job was cleaning houses, just like my mother. She liked the money and independence her work provided. I made sure she was in my clean team as much as I could. She also had a lot of glitter. She really wants to go to the burn, and I hope one day we can meet there. The cool thing about working in a music festival is that you can plan your cleaning route around the music sets that you want to attend and take your break accordingly. We could even drink during the shift, I think. Even though we worked, there was still a lot of time to enjoy the festival. I didn't know what I'd do if I had more free time. There were also some perks about being "crew" such as going into areas that normal festival goers couldn't and, most importantly, the crew bars which had better priced drinks.

When Glastonbury started, we all worked crazy amount of hours, switching the full shit bins, troubleshooting and cleaning toilets. To swap the bin was a real workout. If the urine didn't drain properly than you'd be moving urine which is heavy. One liter of water weighs one kilogram. Those bins could take 200 liters and normally weighed as much as 100 kilograms. Given the shape of the bin, tall and narrow, in order to move it, you had to engage it with your upper body and "dance" with this shit filled bin, moving it back and forth out from underneath the toilet block. But before you could dance with shit, you had to unscrew the urine hose which released any urine that had not been drained.

Although, I was apprehensive about not being able to practice swimming leading up to the festival, at least I took comfort in the fact that I was exercising my upper body which is the more important part for swimming. At first, I used gloves but after a couple of days, I literally stopped giving a shit about shit and just casually unscrewed the hose with my bare hands. At least we had a lot of hand sanitizer available. We then lined up all the shit-filled bins in a double row so the "Shit Muncher" truck could pick them up, or much them. This truck was a normal municipal trash collector truck that violently lifted bins up out of the ground tilting the shit into the trash bay. I can still remember the brown shit falling into the truck and smearing the green plastic bin walls. Then, we had pick up empty clean bins that had been previously left there, or not, and place them under the toilet and screw the urine hose again. All the while, someone from our team had to go in front of the toilet and close it so that no one took a shit while there was no bin underneath, although that did happen a couple of times. People just don't look, or care, where their shit is going. Later, we'd have to spray the shit smeared green bins with a lot of water and detergent.

There was one moment that I will forever cherish in my Glastonbury shitty memory bank. There was a huge toilet block with more than 50 toilets next to the main stage called Pyramid stage. There had been a call in the radio that this block was extremely full and needed to be swapped out ASAP. It was a shit emergency. Some of us went on foot and arrived to huge lines with angry festivalgoers who looked like they wanted to shit, badly. Most of the toilets were filled beyond capacity; shit was literally flowing out of them. We started closing the first units but then more of our colleagues showed up in what felt like the cavalry heroically arriving to the rescue while "The XX" started playing their *Intro Song*. It was magical. We all banded together like soldiers and swapped all the bins filled to the brim with shit and piss. Every time I listen to *The XX*, I think back to this day with nostalgia. It's amazing how we bonded over fighting the festival goers and their shit during that week. Since there had been such a big build up leading to Glastonbury, we were all relieved when it was over on Sunday.

Then, it became grim, real grim. On Monday, we were given the day off to recover. As we walked back to our camp on Monday morning, the festival looked like *no-man's-land*. It was as if a bomb had gone off. There were abandoned and broken tents everywhere. Cyder cans both filled and empty were thrown all over the grass. Cheap Decathlon camping gear was left behind. Everything and anything that you could imagine was up for grabs. Most of these festival goers, waited for years to attend the festival, they had saved up money, bought gear and costumes but now it was all trash they didn't care to bring back or perhaps took so many drugs that they couldn't even make it back to their tents. It became just a depressing sight, symbolizing waste, wealth and privilege.

I joined my Hungarian co-workers who had been to Glastonbury before in one of its traditions called "tatting," which consists of going up to abandoned tents looking for valuables. I found a nice pair of boots, a great 60-liter backpack, loads of cider cans, clothes, money and more drugs. One medicament I kept finding in most tents was used to deliberately give the user constipation, *Imodium* which is a drug used to treat diarrhea. This means that these people without diarrhea thought the toilet experience was unbearable to the extent they'd rather give themselves a constipation than to take a shit. When you consider that most of them were on drugs to begin with, and hence could benefit from excreting toxins from their body, it left me speechless.

After Glastonbury, I was able to escape the camp to train at the local 25-meter pool for two sessions. In July, I got an email from Rafael's account, but it was written by Laura, his daughter. She was informing me that her father was sick, all the swimming slots for September were filled up and that I was at the back of the line. "Not again!", I thought. I told Balú, one of the Hungarian mates about the swim and he called me crazy in his cartoonish accent. Throughout the summer I tried to find pools close to the festival sites, but it was nearly impossible. In the three months, I could only practice about five times.

When you're working at a festival, you're called "crew," and if you're lucky, there is food catering at the festival which your employer has paid for. Thanks to NE, at Glastonbury, we had this luxury. Catering reminded me of American High School cafeteria, wood benches and all. On one normal shitty day of disassembling toilets, during my lunch break, I spotted a girl with a jacket that read: "Mexico is the Shit." They have their own Instagram account and accompanying hashtag #mexicoistheshit. I almost flipped. I had just come from México, and here I was looking at this jacket. I couldn't resist so I went over and talked to the jacket owner who told me they were launching this jacket. The jacket was pricey, and I hoped they could sponsor me. I would gladly promote their jacket. I sent them a message telling them about the toilet in Oaxaca and my work in

Glastonbury, but they weren't impressed. I just thought of this jacket as a cool sign from the shit telling me I was on the right path. One day they will talk to me.

When I lived in South Florida, I trained with two United States Masters Swimming, USMS, teams. First, I trained with Hammerheads out of Davie, FL led by coach John who is undoubtedly one of the best coaches that I've had the pleasure and pain of training with. He always pushed us hard. It's as if he knew we still had more inside of us which he wanted to see in the pool. He was old-school in his tactics and I appreciated it. He spoke his mind and openly criticized my performance because he knew I could swim faster. I am grateful for his coaching. I became more confident of my swimming under his training and registered for my first swimming meet, as the local races are called. When I was registering online, I had to select our team from a drop-down menu, and I was surprised to read "Hammerheads-LGBT Chapter." I had been training with a gay swimming team for over one year and had no idea.

Then when I moved to Miami to work with Bouygues, I looked

for a pool close to my apartment in downtown and was happy to find another USMS team called Nadadores which is Spanish for Swimmers. They were also a LGBT team but this time I knew it going in. It's Miami, what do you expect? The coach is an Argentinean expat but we also self-coached practices (Thanks Doug). During the week, there were two evening practices and weekend practices were Saturday on the pool and Sunday on the beach. I like swimming with faster swimmers. Usually, we organize ourselves in a pecking order with fastest ones on one side and the slowest on the opposite. I tended to swim in the middle of the pool, not so far from the fastest swimmers. I liked to swim next to them and every time we swam past, I could observe and learn from their technique. When there are many swimmers in a pool, you will rarely get individual attention from the coach, so you must take responsibility for your improvement by copying what the fast swimmers do. There was a swimmer named Trish who was among the fastest in the pool. She had strong arms and perfect streamline; she looked effortless. Swimming can be a beautiful sport, bodies gliding in the dark blue waters. She looked as fast as a dolphin.

Faith is an incredible thing and I liked to remember my triathlon in Key West. Even though I wasn't sure that my performance was being tracked by the race organizers, I still raced my best and I was going to train for Gibraltar even though I had no idea if I would be able to attempt the swim. Have you ever trained for a race in which you didn't know you would participate? However, I was completely involved in the toilet work during the summer and finding a pool nearby was becoming nearly impossible because the festivals tended to be isolated from the cities and getting in and out of the festival grounds was a hassle. I tried a couple of runs and biked inside the festival to get to the toilets as a way to keep fit. But these sporadic swims would not be not enough.

At the end of the summer I was contacted by a human resources agent on behalf of an up-and-coming European start up hiring for their San Francisco office. They were manufacturing a water-saving shower, the shower astronauts will use in Mars they said. We had a couple of phone interviews and they asked me how much I'd like to earn in order to go work for them. The job was going to be training technicians on how to install these showers in luxurious homes and hotels. They liked that I was an engineer and spoke Spanish. There would be a lot of travel involved. Yet, there was no shit in the job description. I gave them a really high salary because I felt that whatever price I gave them would be like putting a price tag on my freedom. I asked for 150,000 dollars. Living in San Francisco is expensive, but I really didn't want the job. I had found my passion and it wasn't showers. They didn't make me an offer.

Days later, I was mindlessly scrolling through Facebook when a Brazilian start-up founder who I started following weeks prior posted that they were looking for a Portuguese-English translator for their online bookstore. The job consisted of translating 200 book summaries from Portuguese into English. In my early days of college, I took a translation and interpretation certification class and had actually considered the profession given the amount of travels and my proclivity with language. However, I didn't pursue it wholeheartedly; instead I did freelance work occasionally. Seizing this opportunity, I messaged the CEO and I snatched the job. I was going to be translating non-fiction book now that my summer toilet work was ending. Little did I know how crucial that work would become.

7-SWIMMING
IN TARIFA

I was happy the summer was over. I spent a couple of days in London to have a proper shower before boarding a plane to Toulouse where I was meeting a girl who I consider my little sister, "Hermanita Mariana." She drove us to see my friends Arnaud and Alizée getting married in Béziers, southern France. He had been my first Frenchie friend and the first person who inspired me to learn French. We've kept in touch since Buenos Aires and I was honored to be invited to his wedding with lovely Alizé. The reception was in a wine farm deep inside the country and we had a great time.

On September 4th, Marianne and I flew to Malaga from Toulouse. It was invaluable to count on her presence at that time. She organized all the logistics by requesting Couchsurfing accommodation, booking the buses and researching the cities we were about to visit. I just love this girl. Her dry humor, her openness and ability to deal with change made her my best travel partner. I admit, I wasn't open to traveling with many of my friends but given her previous experience hitchhiking to Alaska from Montreal with our mutual friend Anastasia, I trusted her skills. She took care of the trip so I could shift my attention to training. She's an artist, and we had talked about her illustrating the swim with her drawings. It was going to be great, we thought.

The first thing she did was get us a couch with Laurence, a Brit expat living in *La Línea* which is the Spanish city steps across the border with Gibraltar, the U.K. territory where he worked during the day. Counterintuitively, the Gibraltar strait swim doesn't start from Gibraltar, but from Tarifa. Yet, Gibraltar is a must stop on the way there and Laurence was an incredibly nice host who became a good friend. We spent one day to tour Gibraltar with its imponent rock, monkeys and fastest border crossing. I went from the Spanish side to the British one in 15 seconds. As an engineer, the most interesting thing about Gibraltar was getting to cross the one-runway airport by foot after crossing the border. At night we went to a beach bar where I talked to an Irish expat about my swimming plans while witnessing drug smugglers darting across the sand carrying hashish-filled boxes recently brought by boat from Morocco onto fast-speed motorcycles for the European market. Morocco is the world's greatest exporter of the drug made by compressing the marijuana plant and extracting its resin. After looking at the crime being committed in front of us, the Irish guy wondered how I was going to swim across the strait given the fact I hadn't trained much. I wondered about that too.

Next morning, we left Laurence's place and took a gorgeous bus ride into Tarifa. The bus offers a spectacular view as you follow the rugged sun coast of Spain, along the edge of the European continent, overlooking the Strait of Gibraltar deep down and Morocco on the other side. When I first saw my swimming destination, I felt that cold sensation in my stomach as blood rushed away from it. As we got off the bus, we were welcomed by a warm breeze of salty air. Tarifa in September is terrific; it's still sunny and most of the tourists have left. We headed straight to the swimming association's office to meet Laura and ask her if I could swim across the strait. Laura opened the glass doors, welcomed us with a tense smile and proceeded to lash out reality to me. Marianne speaks Spanish, and she understood

Laura telling me that 2017 had been the worst year for crossings with less than one-third of athletes being able to attempt the swim as the year before. To make my case worse, I had not been given a swim window, and I would be joining the back of the line at number #23 or #37, I didn't matter. Finally, her father Rafael had passed away the previous week. I felt powerless to console her, all I could do was to look her in the eyes.

The swim season would end on October 30th, seven weeks away, she alerted me. Laura then showed us the ideal swimming line of 15,400 meters connecting Isla Paloma in Spain to Punta Cires, a rock outcropping off the coast of Morocco. There were pictures of previous crossings hanging on her office wall displaying proud swimmers standing on Moroccan rocks. I was anxious but I was not surprised; I looked at her in silence hoping for a lifesaver. She said I was welcome to stay in Tarifa without any promises of swimming, and in the slim case there was a spot open, and no one from the queue claimed it, then I would be able to attempt the crossing. I took a deep breath and said, that for me, that was enough; I would stay. Laura then exhaled, sighing in relief, and showed us some cheap eats, local attractions and suggested we both go to Morocco for a weekend, after all, it was only a 30-minute ferry ride away.

We left the office, and Marianne rightfully wanted to pull my head off. She screamed in disbelief: "*ALDO!!!*" with her cute French accent. "How can you fly us here without being allowed to swim?" She had worked on a communications plan which I now wouldn't be able to follow since I didn't know my swim window. I explained to her that I had been trying for the last two years and they always asked me to come back next year. I had to come here since I was in Europe that summer. I couldn't waste this chance, I tried to reason with her. She was also angry because we had planned for me to swim in one week then go to Morocco at which point, she would fly back to Toulouse where she would start a new job. I tried to calm her down by suggesting we walk to the *mirador* which is a famous sightseeing spot in

Tarifa from where you can see Morocco.

The view from the *mirador* was breathtaking. On the Moroccan coastline, there is a pronounced mountain called Yebel and during my time in Tarifa, every time I spotted it in between the clouds, I felt as if Yebel was inviting, almost daring me to swim towards it. I would always look at it and say: "Yebel, I'm coming." But after realizing that I wasn't going anytime soon, Marianne went back into planning mode. She started looking for her best way to get back to France while we made the best of her week with me. She started illustrating under an umbrella of the windswept beach while I went to the ocean for the first time. But getting in the water was another big slap in the face. First, it was cold. I think the water was about 18°C (56F) which shouldn't be cold, but it felt very cold to me, being used to Miami water temperatures of 30°C (86F). Then, I barely managed to get a few strokes in before the waves spat me out of the ocean onto the sand. I felt like the Strait was showing me who's the boss. At this point, I felt incredibly inadequate in that town. I had had enough and doubtfully walked back to check on Marianne. She was much calmer now, smiling like a child on her first beach holiday, drawing the sand and the waves.

Tarifa, 1st day of training

(Drawing by
Marianne Bruyeres)

Back in Florida, my mother, uncles, and friends, were bracing for another hurricane. This time it was called Irma and it was projected to be a destructive category five. Hurricanes range from category one to five depending on their speed and consequential damage. A category five destroys houses, buildings, cause floods, and power outages for days. Great, I thought. After leaving in Florida for 13 years I had learned to cope with the stress of Hurricane season, but it was not what I wanted prior to starting my training. I said a prayer to them and hoped the hurricane would turn away from Florida and its wind forces lessen. The main concern when preparing for Hurricanes is to stockpile water bottles and I wondered if they would also question how we use it to flush our feces every day.

I went to Decathlon which is a chain of sports accessories stores in Europe and bought clear, oversize goggles. Pool swimmers usually wear small sized glasses to reduce drag on the water, but I wanted to see as much of the ocean life as possible. I figured

I wanted to enjoy the scenery during my training. I also don't like the optical illusions created by small goggles on my peripheral vision where shadows look like sharks. Perhaps, it's time for me to disclose a big fear of mine. I was afraid of swimming in open waters. I'm not so much scared of sharks, even though once I bought a device called sharkbanz which is just a magnet said to repel sharks. Waste of money. Every year, 0.5 persons die from shark attacks according to the New York Times. Fear is an irrational behavior, and I always conjured up marine monsters in my mind. I think this is one of the reasons I like ocean swimming. I wanted to face and eventually overcome this fear too. In Miami Beach, it was reassuring to swim with others, but now, I'd be training by myself. I got these windshield-sized goggles and went back to test them in Playa Chica, Tarifa's small beach.

Once after a tropical storm in Miami, millions of jellyfish were brought closer to shore where I used to train in Key Biscayne. It was a surreal experience to swim with them, it was like floating in space with aliens. When I was about to go in the water, a swimmer coming out told me it was ok, and that I would unavoidably hit them but that their sting would go away. Jellyfish are really the biggest deterrent to open water swimmers, not sharks. They are slow moving, show up from anywhere and their burn hurts, sometimes forcing us to abandon the crossing. Luckily, it was not season and I didn't see any jellyfish during my first swims.

At this point, I still needed a place to live. None of the active Couchsurfers answered my couch requests for crashing with them. Tarifa is a popular tourist destination in Europe, famous for being one of the best spots in mainland Europe for kitesurfing and its geographical location as the southernmost point in mainland Europe. In the two months I was there, I met many kiters who come from all over the world to enjoy the wind.

Since I couldn't get a couch, I searched Airbnb and found an Austrian guy called Alex renting a room. We talked on the phone, and I agreed to see him later in the day. My first impression of

Alex was that of a "Papa Bear." He's tall, smiley and immensely endearing. He had been returning to Tarifa on his yearly summer migration. Many kite surfers are what we call digital nomads, that is, people who work from their laptops and travel to exotic places. After meeting Alex, I was looking forward to sharing the flat he was renting on Airbnb and learning about the digital nomad lifestyle. Remember the translation work? I had more than 200 books to translate, and I organized my schedule so that I could train in the morning and/or evening and work during the day. I was also going to become a digital nomad.

Things seemed to be falling into place, and I tried not to be apprehensive and enjoy the training ahead of me. Luckily, Tarifa proved to be the perfect place for work and fun. Alex introduced me to his nomad network, and I fit right in. Digital nomads, to me, are ambitious people who don't want to compromise on life. They were all incredibly hardworking and, yet, knew how to have a good time. That's why most of them did some type of extreme sport such as kiting.

There is a Facebook group called "Tarifa Digital Nomads" full of activities geared towards the community, and I joined as many as I could. There were breakfasts hosted by Manon, a Dutchie living there, hikes, talks, and trips planned by passionate nomads. Tarifa was looking like the place in which I wanted to be. I also enjoyed the uniqueness of being the only person training to swim across the channel. Unavoidably, the conversations in Tarifa will always turn to the wind and surf conditions, so it was refreshing to talk about something different: swimming for toilets!

A couple of days into my lonely training, I spotted a group swimming in line. They were wearing bodysuits (neoprene), and I went to join them. It's not so fun swimming by yourself for a long time. I ran on the beach and tried to catch up with them, a sad attempt on my part. They were fast swimmers and only at their halfway point was I able to meet them. We turned back, and I was left in their wake again. At the beach, I learned they were training for an attempt later that weekend. The swimmers asked me when my swim window was. I said I didn't know it. They looked surprised. The swimming window is the period between seven to ten days where you're asked to come to Tarifa and, weather permitting, will be allowed to attempt the crossing. Weather permitting is the key word here. Remember that Tarifa is famous among kite surfers? The reason Tarifa is called "the wind capital" of Europe is that the Strait of Gibraltar acts like a funnel intensifying the wind currents in the region. There was always talk of which wind was prevailing: *"Poniente"* or *"Levante"*, which is Spanish for "westerly" or "easterly", respectively.

I came to love/hate the website Windguru because I'd start my days by checking the weather and surf conditions. Unlike my kitesurfing friends, I was always hoping for no to little wind which would in turn not generate big waves and allow for channel crossings. This is a sample screenshot of the wind chart.

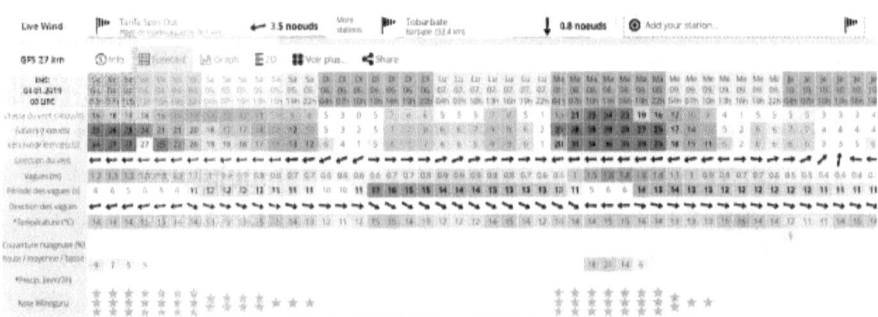

Source: windguru.com on Sept 4

It's frustrating to follow the wind, I realized. It's always changing! I didn't care so much about the wind direction as much as the wind intensity. I also profoundly cared about water temperature and wave sizes. Winds less than 5 knots would be ideal, generating waves less than 1 meter tall and the water temperature in September was around 19°C. My ideal graph color was white to turquoise, and the worst case would be red, meaning strong winds. As mentioned before, winds generate waves. Stronger the winds resulted in taller waves.

My favorite training spot became Playa Chica. It's protected on one side by a road connecting Tarifa to Isla de Palomas and the port of Tarifa on the left side. One lap was about 650 meters, and the first time I swam it, it took me 17 minutes. It was not encouraging. Laura told me I needed to swim at least 3 kilometers per hour to reach Morocco and this pace was less than 2.5 kmph. She told me that if I couldn't keep the required pace, then I wouldn't be strong enough to overcome the current in Morocco and no swimmer had accomplished the swim at a speed below 3kmph. I realized I needed at least one month to build up my pace and changed my mindset to train and enjoy my life in Tarifa. More time was actually a blessing and I became grateful for not being able to swim at first.

On the 15th of September, I aimed to stay in the water for one hour, but the cold was destroying my mind game. I just had never swum in waters at this temperature. I started thinking about getting a wetsuit. Later that night, I went to a concert a Tarifa Eco Center where Manon was working and was a favorite spot among nomads. There I met Florian, a Swiss guy, enjoying his unending vacation months. I had found him on Couchsurfing and, thought I had sent a request, but I never heard from him. I immediately recognized his cycling cap or "casquette" similar to my friends' from the Miami Bike scene. He revealed that he had not received any requests from me. Weird, I thought. Perhaps I just assumed that he wasn't going to reply and never sent him a message. I explained to him that I was training to cross the channel, but the cold waters were getting to me, and he kindly offered me his suit.

Next day, I went back to Playa Chica and managed to swim 46 minutes at about 1.7kmph. Flor's suit was a half-body suit for kiting and thick which is horrible for swimming, but it kept me warm. After that morning's training, I met an Argentinean living in Tarifa at the beach shower rinsing away the saltwater. He told me he had made the crossing and was swimming as part of his morning exercise routine. He encouraged me and wished me luck to cross the channel. But before leaving, he told me to conserve energy for the last hour. He said I needed to swim fast to break across the Moroccan current at which point I'd be tired from hours of swimming. I was thankful for the tip, but it made me the more worried. If couldn't swim at 3kmpr for two hours, then how would I have remaining energy to break that current?

Days later, I went in the water at 9 a.m. without the suit and tried to fight the cold mentally. I ignored it, and I felt as if my body was making a heat shield of my skin. When the water is cold, you get a burning sensation which is actually heat leaving the body. I swam for 27 minutes at 1.4kmph. There would be no way for me to cross the channel at this pace. I also noticed having the sun shining on me improved my spirits and I hoped for a clear sky day and I also decided to buy an appropriate suit for training.

All the training was getting to me and I started feeling a right knee pain that I knew from my triathlon runs. I had been diagnosed with a right meniscus tear. The meniscus is a cartilage that absorbs impact from the bones above and below the knee. In addition, the pain was also caused because of what I had been told by a sports therapist was Iliotibial Band Syndrome. The IT band is a tendon that goes from your pelvis to your knee, connecting everything on the way, from glutes to thighs and tibia. This syndrome happens when the band is too tight, and it impacts the proper movement of the knee causing pain and discomfort while running and, in my case, also swimming.

On September 19th, after watching a YouTube video about "IT band syndrome massaging." I followed the recommendations of pressing my thigh against a tennis ball which works like a deep tissue massage. The massage did wonders, and I swam 2 laps at 2.2 kmph each at Playa Chica. When I came back to the apartment, I found out that an earthquake had just taken place in Mexico. I contacted my closest friends in the capital and was relieved to hear they were safe. Only old buildings built prior to new construction code collapsed. Thirty-two years prior, on the same 19th of September of 1985, México suffered a similar earthquake which killed thousands. The authorities then updated the building code with more stringent regulations and all structures built in compliance with the new code withstood the shake. However, older buildings which didn't collapse in 1986 didn't resist the seismic waves this time. It's interesting to note that earthquakes propagate in waves, like water, only structures above the ground are impacted by the waves. Subways and underground remained intact. The images of old buildings swaying, and collapsing were alarming. But it was also touching to see the Mexican people coming together, forming rescue teams, unlocking Wi-Fi networks so everyone could use the internet to share information and find missing ones.

Next day, I swam a bit faster at 2.2 kmph but only one lap because *Levante* was powerful, and the waves were tossing me like dirty clothes inside a laundry machine. Swimming like this makes my breakfast come back up. I only swam one lap, 15 minutes. My Spanish diet was changing at this point. When I first arrived, I couldn't get enough of Jamón Serrano. Hmmm, that was so good. I'd buy fresh baguettes from the convenience store downstairs from my apartment and butter it up. The hot baguette melted the butter and I loved the ham's fat. I'd eat this after my swims sometime before 11am. But one month into this diet and I started changing into a more vegetarian based nutrition.

I've been slowly transitioning to a more plant-based diet because of both inhumane treatment of animals and also because of the lower carbon footprint on the environment. As someone who grew up in Brazil, eating meat was an important part of my culture even identity. Nonetheless, I've grown to understand that we don't need, and we shouldn't eat meat every day for both health reasons and care for our planet. I' had increased my vegetarian meal percentage to 90% of my diet and it felt great. In 2020, I went full vegetarian.

My pre-swim meal was cereal with water and fruits, usually bananas or apples. I eat my cereal with water because one swim coach once told me not to drink milk before practice as it ferments in the stomach. I consider nutrition to be the first discipline of triathlon. The sport really is: "eat, swim, bike and run." Food is fuel, I've come to internalize. I think it's also the hardest part of my training. Over the years, I've developed some habits which helped me be more disciplined. However, I still needed a lot of practice in eating right and I was looking at this swim as a way to help me. I began to slowly eat less *Jamón Serrano*, and before I left Spain, I could barely look at it. I wanted to get proteins from more vegetables and eggs for their simplicity of cooking. I also ate a lot of fruits throughout the day. When I went shopping, I always bought bananas, apples, oranges and a local variety of fruits. At this time, I was not creative when it came to vegetables in the kitchen, so I'd buy carrots, tomatoes, red peppers, onions, garlic, and try to make something edible out of them.

Once I had an easy swim earlier in the day, but that didn't mean I was passive in Tarifa. After lunch, I joined Ryan, a kite surfer, who let me use his kite and learn the first lesson of kiting: how to fly a kite. I thought that my childhood experience of flying paper kites would help, and I was partly right, except for the fact these kites are extremely powerful. I was a confident beginner and did well for the first minutes until I flew it into the power zone. This zone is where the wind is strongest, and the kite picks you up from the floor and drags you for one hundred meters along the sand until your whole body is rashed and bruised. That happened to me twice before I decided to stop my kite lessons and keep myself from getting an injury that would prevent my crossing from taking place. However, this small taste of kitesurfing adrenaline made me want to come back to Tarifa just to learn how to kite. After this scary lesson, we went to a favorite spot called *Ion Club* where hundreds of surfers hang out and relax after a day in the ocean.

Next day, I walked 8.5 km (5.3 miles) to Ellen's house on the hills outside of Tarifa on what felt like a portion of the famous "El Camino de Santiago." She's a Dutch digital nomad who devotes her life to helping entrepreneurs start and advance their business by living out of trust, instead of fear. Alex introduced me to her, and we'd always run into each other. She lived and traveled around Europe with her dog Puka but had settled in Tarifa for a while. In front of her house, there was a tree with a swing, and it was also there where she held Tertúlia meetings which were bi-monthly gatherings where people shared art, talks, and food. I participated in one such meeting where she walked with us through the surrounding hills while describing the main steps of the Hero's Journey. She gave invaluable advice, one of which is completing this book. Before I left her house, she invited me to speak at the next Tertúlia and share my toilet journey. Ellen got me in touch with two special people: Thomas and Bianca, who I will talk about later. The walk had been worth every step.

But while I was walking back to Tarifa, the earth was shaking again in Mexico. This time, on September 23rd, Ixtepec, Oaxaca felt a magnitude 6.3 earthquake. I immediately thought of my friends from Heifer and contacted them to check if they were safe. Thankfully, they were all fine, but they reported to me that Heifer members living in the Isthmus of Ixtepec had felt the brunt of the shake and that many houses had been destroyed. Sadly, Mexico and the world were getting earthquake news fatigue at this point and a lot of the media coverage centered on Mexico City while Oaxaca was forgotten.

When I donated the first urine diverting toilet to the family in Oaxaca, I had coordinated with a Heifer employee named Rosario. She was excited about the toilet and we had agreed to follow up on this toilet with more in the future, once we had tested and the family was using the toilet adequately. When the earthquake hit, I suggested that I use the swim to fundraise more toilets to the survivors in Ixtepec and Rosario agreed. I started planning the fundraising and created an account on *Gofundme*, a fundraising platform.

The next day, I diversified my training a bit. In the morning I went hiking with some of the nomads on the sand dunes of Boloña and in the afternoon I went horseback riding on the beach. The day before, when I spoke with Ellen, I had told her that one of my dreams was riding a horse on the beach. When she heard that, she motioned me to stop talking and contacted a Dutch friend called Bianca who keeps her horses in a stable, pretty much on the beach near Tarifa. Within 26 hours, I was galloping on a beautiful white Spanish war horse called Empiria on the windswept sands of the Gibraltar Strait next to Bianca and her dog named Guapa. I was grateful to her for trusting me.

If you know any horse owner, they will not let any stranger ride their horses because there are many risks associated with horse-back riding. I told her about my experience level "near cowboy," and we went for a beautiful and gentle ride. To finish this fantastic day, Laurence, the Brit who hosted Marianne and I in La Línea, came down to Tarifa and we went out. Next day was going to be Sunday and also my rest day.

I've learned that rest days are equally important to training as active days. The body needs time to repair the muscle tissues which were torn during the training. Foods rich in protein help to recover the muscle fibers as well. Furthermore, excessive training can lead to fatigue and injury. I've also noticed that if I plan an ambitious training schedule, then I won't appreciate my training and I will undoubtedly fall short and feel discouraged. I have also learned from reading triathlon books not to try any-thing new on race day. For example, don't try new foods or new gear. And once I realized I couldn't fight the cold, I decided to swim with a suit and improve my chances of finishing the swim. Swimming in Florida, I had never needed one, and they can get quite expensive. When I went back to Decathlon, I was appre-hensive of the low-quality suit I could afford. They only had one inexpensive suit, 60 euros and specific for swimming. But It was so cheap that it didn't specify the millimeter thickness, usually 3 mm are suitable for swimming.

I got back to Tarifa by 20:30 and it was already dark, but I really wanted to try the wetsuit. Have I mentioned I'm afraid of swimming in the ocean at night? Even though I knew Playa Chica well, the thought of going in pitch black water frightened me. So, I did what I often do; I put myself in a scary situation to see how I would react. When I wore the suit, I felt like super-man. Even the act and sound of zippering the suit gave me a confidence boost. It signaled that I was ready. I felt as if my chest was tightened into an armor. No jellyfish could sting me. When I walked in the cold water, water infiltrated the suit, but it stayed in place allowing my body to warm it. The suit also helped me float.

The lights from the street could still reach the beach, and I started treading on water with my head above the surface as I was too scared of looking into the black water. I went farther, 50 meters from the shore, where the lights were dimmer. My heart was pumping; I was so scared. What if an Orca came and ate me? That was impossible, I reasoned. That would also be the first time an Orca attacked a person, ever, but I didn't want to be the unlucky first. There are actually no Orcas in Tarifa in September. They do swim by in midsummer, between March and July. But what if there was one Orca swimming late, and hungry? Don't be ridiculous, I told myself.

I kept going in. Darkness is the absence of light, I tried to calm myself. When I was perhaps 100 meters away from the shore, and it was black all around me, I put my head in the water and started swimming properly. I started by looking forward to my hands and it was then that I realized that every time my hand moved, there were lights in the air bubbles, like glitter. I observed it closer and realized it was bioluminescent plankton! I was swimming in bioluminescent plankton in Tarifa! That was insane. How come nobody knew about this? I wondered. I guess not many people go in the water at that time. I was elated, look at what happened when I faced my fears. I enjoyed it a bit more and swam back mesmerized by my arms moving in front of me creating glittering *bolhinhas*. The suit had passed its first test and showed me that if I was brave enough then I would witness amazing experiences.

Two days later, I was getting ready to start my lap at Playa Chica. I had my feet already in the water and as I was setting my watch's chronometer, I noticed an adult German shepherd circling me with a tennis ball in its mouth. When I put on my earplugs, swimming cap, and goggle, I get in the zone quickly because all sounds get muffled. At this point, I was so focused that I ignored the dog bumping into me, wanting to play and started swimming to Isla Paloma, 325 meters away.

About halfway into my swim, I feel something touching my feet. No fish will bump into a swimmer, so I was curious to see what was behind me. When I looked back, the same German shepherd from the beach was frantically clawing me, trying to stay afloat. I couldn't believe it. The dog had followed me in the water for more than 150 meters. The ocean was choppy, and I could tell he was not having a good time. I'm sure you know this, but dogs' paws have claws. I didn't know this, but they stuck their claws out when they're swimming. And those claws were knifing my brand-new suit, but I couldn't care about it at the time. All I could think was "I'm glad it's not my skin".

I had to think fast, but I didn't have many options. Since we were about midway, to swim back would be as long as reaching Isla Paloma, but the way I was getting clawed signaled the dog couldn't swim either way. There was only one choice: to swim into the rocks and climb onto the bridge that connects Tarifa to the Island. I was not looking forward to this because the waves would toss us both into the boulders which are both slippery and have razor sharp corals. Still, I had no better alternative, so I started paddling towards the rocks with one leg and kicking with one foot while lifting my right arm and right leg so the dog could support itself and stop clawing my brand-new suit and one-and-only original skin.

As this happened, a couple of people started gathering on the bridge to watch this impromptu rescue. I must have been 50 meters from the rocks, and once we arrived, I waited for a break in the waves sequence and pushed the dog towards the rocks. It then realized what I was doing and started paddling. The waves came and slammed us onto the rocks and then washed us back down and we managed to grab onto the stones, sliding, scratching but holding on. We both climbed up the rocks wall, and I realized I had survived my first dog attack in the water, second overall. Curiously, I was attacked by my family's German shepherd when I was an infant. It's so early that I don't remember. But I do know that my father put the dog down after they found me lying in a pool of blood. The beach dog owner, a stocky lady, came running with her leash in hand. She desperately thanked me, but I was then able to think about the suit that I rushed to the shower to wash the salt and check for damages. There were superficial scratches on the back of my legs, but nothing major. Thanks, cheap decathlon suit for saving my skin!

The next days, I managed to swim 45 minutes at 3kmph. My first week's efforts were starting to show results. I was slowly adjusting to Tarifa, working on my translations and meeting inspiring people. I was increasingly happier. I checked the wind, and this is what I saw on the 26th.

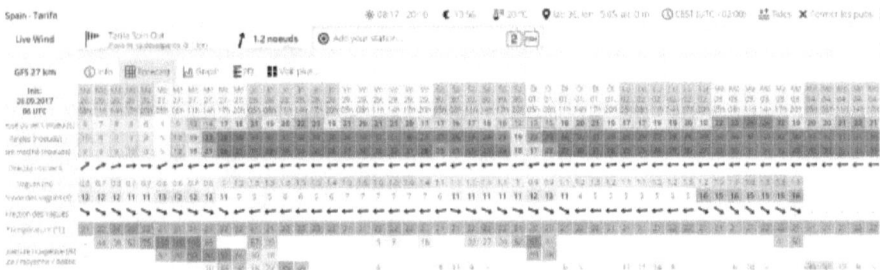

Just look at the first row with numbers: "8, 7, 8,8..." The left column says "...esse du vent(Noueds)" which is a truncated vitesse, or speed, of the wind in French. Noueds are Knots which is the unit for windspeed. 1 knot equals 1.85 km/hr. or 1.15 miles per hour. Remember, I was looking for light colors, white or turquoise? Yellow and red meant strong winds where no one was allowed to swim.

My kitesurfing friends were in for a good week though. All I could do was wait and train harder. Once I trained in the morning, worked on the translations in the afternoon and then met Christian F. He was one of the Couchsurfing members who had not replied. He was back in town, and we agreed to meet at Café Surla whose owners are his friends. Christian is one of the people that make Tarifa such a special place. He had toured the globe on motorcycle and settled in Tarifa where he was working on ecological projects. He had also swum across the Strait with three friends years earlier. I wanted to meet him to learn about his experience, get advice and learn about his travels. He also gave me the same advice as the Argentinean swimmer: save energy for the end of the swim because once I'd approach the Moroccan coast, I would have to break the current which would otherwise take me into the Mediterranean Sea. He also lent me his GoPro camera to record my training.

After speaking with Christian, I went outside the restaurant to meet Rodrigo and Ana. They were Couchsurfers from Portugal who had arrived in Tarifa the night before and tried to sleep under the wood flooring of a beach bar to avoid any police. It's no surprise they couldn't sleep given all the people dancing above their heads. They had no other option because no one offered them a couch, but they were hardened travelers. They sent me a desperate couchsurfing request and I convinced Alex to let me host them. He sounded skeptical at first, but I told him how couchsurfing worked and he agreed to having them in our spare guest room.

Despite their tiredness, when I went out of Surla to meet them, they became ecstatic and group-hugged me. Rodrigo reminds me of Beaker from the Muppets because of his glasses and how he was always flailing his arms while talking. They were tall and loud, often interrupting one another as their excitement took over, like clumsy giants. They had many bags, more than normal experienced travelers carry and I offered to help. Rodrigo had his long board and set some of the bags on it and we started walking up towards my apartment.

On the way there, they started sharing their adventures hitchhiking to Tarifa. They were speaking fast, and I was not so used to Portuguese from Portugal, so I needed to interrupt both of them and ask for clarification. They sounded funny. The language difference is similar to U.S. versus British English. Sometimes we employ the same word for different meanings. Even though I was overwhelmed with their talking, I realized that the bags they were carrying were actually grocery bags and I asked them why they had so much food to which Rodrigo replied they had gotten it in the dumpster. "The dumpster?!" I asked. The dumpster, they confirmed. They went on to explain to me that they practiced "Dumpster Diving." They'd go behind supermarkets where there are dumpsters and look inside for food, as simple as that. I couldn't believe it, so they invited me for a dive.

We got home, and Ana went to the kitchen to prepare something for us to eat with the food from their previous finds while Rodrigo and I went back out to look for food in the trash. He was on his longboard, and I was on my bike lent from Acro Yoga Ryan. We went downhill avoiding cars, Rodrigo was yelling crazy profanities from Portugal and I felt a warm summer breeze kissing me. It reminded me of my childhood cycling in Brazil. I was happy.

We rode to the closest supermarket on Calle San Sebastián, six blocks downhill from my place. On the way, Rodrigo stopped at dumpsters plucking their black plastic bags and inspecting them with expertise. I gasped. I was feeling both ecstatic and ashamed. What would my family say if they saw me? But it was an exhilarating thing to do, adrenaline-laden in fact. Rodrigo shouts *"Encontrei maçãs!"* after he finds a bag full of apples, most of which were perfectly fine to eat. Then he exclaimed grossed out "ewww" after some gooey liquid dropped on his foot. *"Porra, caralho!"* he cursed in Portuguese with his Portuguese accent making everything funnier. We both knew what that meant. I was laughing at the scene in front of me, having so much fun in the dumpster that I couldn't believe myself. At the same time, I was embarrassed as the strangers walked by. Were they judging me? Did they think I was a homeless person? But they just walked on and didn't say anything or notice me at all. Rodrigo grabbed another black bag, punched a hole and inspected its contents. If it was wet or dubious, then he would discard it by placing it in the adjacent container so he could continue the process. More pedestrians walked silently behind me. Rodrigo couldn't find anything else, so we skated/cycled to the next dumpster as I asked myself if what we were doing was even legal. *Legal* in Portuguese literally means cool.

We arrived at another supermarket called Sol and went around the corner to its dumpster. Rodrigo propped his longboard against the it and inspected its interior as if it was the most natural thing to do. I stayed a morally safe distance away and tried not to think about the absurdity of what I was doing. On one hand I didn't want to be associated with him, but it was so funny I couldn't help hoping that he would find more food. Bingo, he finds yogurt close to the expiration date and hands it to me saying it's safe to eat.

As I'm eating yogurt from the trash, he had already found more food. He invited me to come closer showing me an opened bag with potatoes, bread, more yogurts, carrots, and even ham. I don't recommend eating meat from dumpsters because meat is not safe to eat. I learned this from Rob Greenfield's dumpster diving guide. He is an expert on the subject, and I recommend following him. Once you get past the first shock barrier, it's not hard to get excited. The food was taken from the shelves, then placed in the bags, so they were clean. There was also a lot of packaged goods one day or two ahead of the expiration date or expiring that same day. There was a lot of good vegetables with minor bruises thrown out. Obviously, this was the dumpster, so there was rotten, spoiled and smelly food too, but if you could separate the bad from the good, there was perfectly edible food being thrown away.

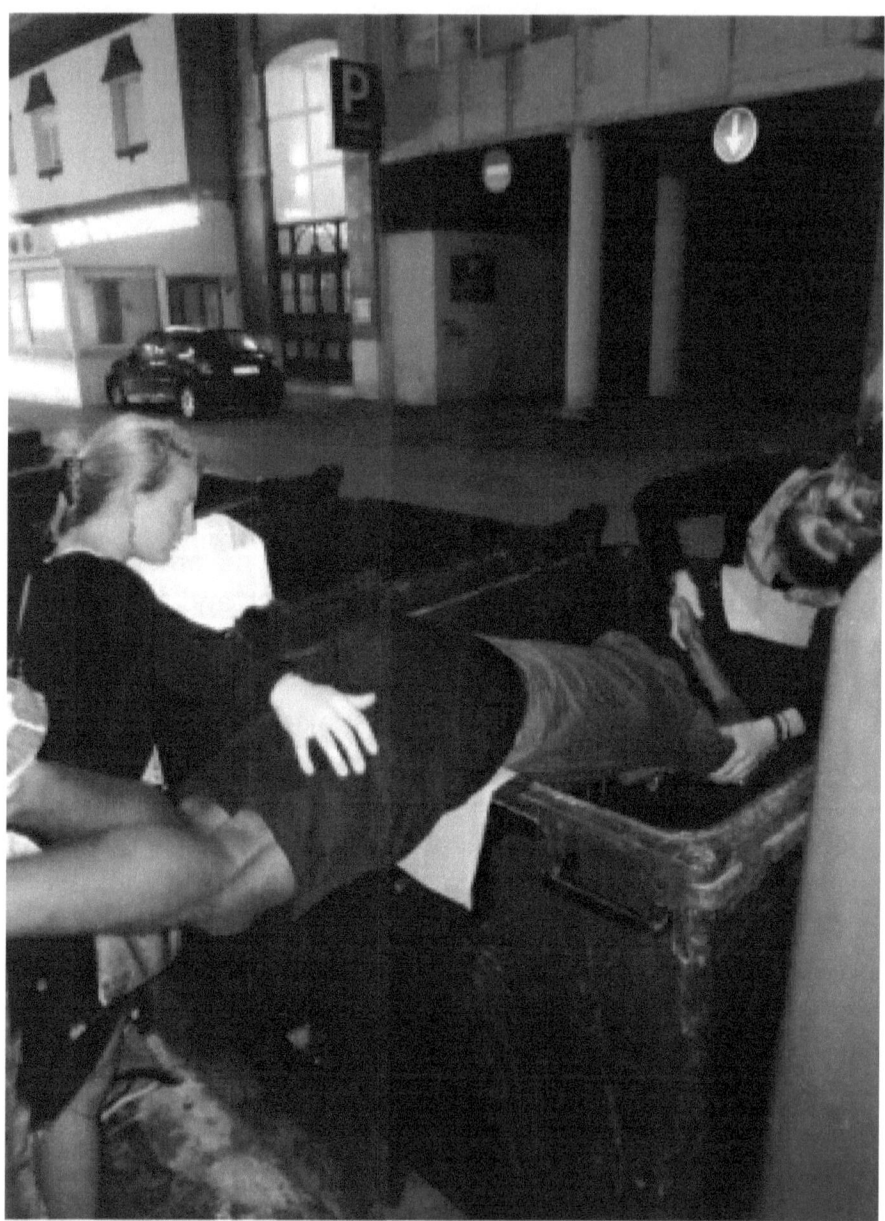

What made food become so cheap that we could throw it away? A lot of what I learned about modern agriculture comes from Michael Pollan. My favorite book of his is called "In Defense of Food." In it, he talks about our relationship with food as not only a fuel but something that creates community between ourselves and the earth. He extolls how much time Italians devote to eating with their families, the health benefits of the Mediterranean and decries how many children have never seen a real chicken in their lives.

Meanwhile, I remembered that in the United States, I learned to eat fast and in front of a television or behind the car wheel, mostly alone. Ironically, I worked mostly at restaurants in the beginning of my university. I had jobs throughout the whole food industry pyramid, from pizza places, cheap burger joints to fine-dining restaurants. In my last years in the hospitality industry, I worked as a meat carver in a Brazilian steakhouse in Miami where at the end of my shift, I could eat leftovers of leg of lamb imported by plane from New Zealand.

The answer to what made food cheap was fuel and energy. After the second World War, there was a surplus of Nitrogen destined to make explosives such as TNT. By then, German chemists Fritz Haber and Carl Bosch had already won a Nobel Peace Prize for developing the Haber-Bosch Process which reacts Nitrogen (N_2) in gas form from the air with Hydrogen (H_2) from natural gas to create Ammonia (NH_3) which can then be used as a fertilizer because the nitrogen in ammonia was in solid form. This process would not have been possible without energy from fossil fuels because it required lots of pressure or heat to achieve these reactions.

After the war, fuels were cheap and there was a lot of nitrogen stockpiled as unused warheads so the mass production of fertilizers and their necessary cousins, pesticides, became possible. In essence, we went from bombing cities to bombing the fields with fertilizers and pesticides. Resultingly, a lot of food was produced and at a cheaper cost. However, not all of the nitrogen from ammonia was captured and as much as half leached into the water. Remember what happens when too much of a resource is available in one place? Too much nitrogen in the water caused eutrophication and dead zones. Aquatic plants and phytoplankton fed with the leached nitrogen explode in growth and multiplied. When these plants and algae die, their decomposition consumes the absorbed oxygen in the water resulting in fish deaths. As more fish die, there is more organic matter decomposing which creates a vicious cycle creating "Dead Zones." Beaches of Miami and Cancún have been infested with these algae which in México they call *Sargazo*. When I was living in Playa del Carmen and when I was swimming around Key West, I saw these sea weeds blanketing the shore, but I didn't realize they were partly caused by excess nitrogen coming from fertilizers. I've had a couple of swims ruined by them and even walking on the beach is unpleasant as they stink when they're breaking down.

Now, instead of famines, thanks to modern agriculture we have a lot of food and subsidized wheat, corn and soy. There is also the advent of monocropping, large swaths of land devoted to the production of a single food. I've driven through some soy fields in my road trips in Brazil and it is a sad event. A few single trees dot the landscape. There are few birds. There is no green, nor diversity. It's silent. Pests arrive precisely because we have created artificial ecosystems. I've learned from my permaculture course that even invasive species have been possible because the top predator has been killed, normally by us. Monocrops farms in Brazil are being attacked by wild herds of hogs brought from Asia. They're invasive because they have no top predator large enough to kill them. I remember that on my uncle's farmhouse there were three hides from the jaguar (*Panthera Onça*) hanging on the wall. If only we had driven the *onças* population to an endangered level, then perhaps they could help us. Instead, cattle ranchers kill *onças* because they attack the calves. If only the *pantheras* were not having their habitat decreased thanks to deforestation to make room for more cattle ranches and monocrops.

In Spain, instead of monocrops there were huge plastic green houses and that's where most of our vegetables in the dumpster came from. They were also only possible thanks to modern agricultural practices. However, Rodrigo seemed oblivious to this as his excitement about what would be in the next bag took over. It reminded me of going up to the toilets at the festivals. All of this with the novelty of doing something for the first time, telling myself there was no shame in it every time someone walked by made this one of the most exciting things, I had done that summer. Plus, this was food that wasn't going to be wasted. It was a win-win for us and the environment; I felt like I was living my sustainability principles. We found more apples, potatoes, grapes and vegetables. There was a mix vegetable container which only defect was a finger-punctured hole on its wrapper. It was thrown away, long before the expiration date. It didn't deserve to be in the trash, but it was and here we were rescuing it from the landfill. It felt good and right to be taking food from the trash. I was hooked.

We hauled our food bags home and came into a house smelling like something good was cooking. Ana had prepared some pasta with vegetables from their previous dive. I was eating food found in the trash. Alex was not keen on trying it, though. After Rodrigo and Ana continued their journey hitchhiking to Morocco, I started dumpster diving on my own and invited my digital nomad friends. I managed to convince two of them, Marina and one who shall not be named as per his request. I read more from Rob Greenfield and every time I pass by a dumpster, I take a look inside. This was one of the reasons I came to love the Couchsurfing community. If you were open to new experiences, you could learn new things which may just change your outlook on life.

On Saturday, I was invited by Ellen to share my toilet journey with her Tertúlia network. I framed it in terms of the hero's journey which we had learned previously, and her friends were incredibly supportive. I was there where I met Alberto, a local kitesurfing, climbing instructor and kayaker. As a fellow athlete, he was moved by my story and decided to help me. He asked how my training was going. I told him I felt I needed to do longer swims outside of Playa Chica but that I felt vulnerable in the water with all the kites and surfers. He then proposed to follow me on my training with his kayak. That would be great, I would be able to do longer swims, and simulate feeding with him.

A couple of days later on Sunday afternoon, October 1st, we trained together for the first time. We met at Playa de Los Lances which was a 10-minute bike ride from my home, on the western side of Tarifa's coastline. It was going to get dark soon, so we did a short two-hour training session, but having Alberto supporting me made me swim faster. It was good to have a different ocean bottom and see bigger fish and marine life. I spotted manta rays and barracuda-looking fish. This contact with nature is one of the main reasons I love adventure sports.

Ellen also introduced me to Thomas Jakell. She met him at the 2016 DNX, Digital Nomad Conference in Portugal where she heard him speak about how he hitchhiked from Berlin to India to crowdfund the construction of compost toilets. How similar were our stories, she remarked. Also, she thought he was from Brazil, and felt she needed to introduce these seemingly Brazilian soul brothers who were doing crazy things for toilets. When I finally caught hold of Thomas, and after reading his impressive story, watching his videos on YouTube, I was thrilled to be talking to him. He had a loud and contagious laughter that will get anyone smiling. He shared with me his efforts and the story of the company he co-founded to provide compost toilets at music festivals, just like I had done in the summer of 2017. I felt an immediate connection with him and thought we needed to talk again. He shared his plans to go to Africa by motorcycle, which would bring him to Tarifa in a few weeks and we agreed to keep in touch and talk again after my crossing.

Alberto's partner Belen happened to be a massage therapist, so we arranged for me to get a deep tissue massage at their place. After weeks of training, I was sore, and a massage sounded great. My life was beautiful, I thought, as I was laying on the floor getting a massage. Here I was, training to swim for toilets while having such a good time. What else could I wish from life? I biked home smiling and incorporated stops at the local supermarket dumpsters so I could have a look inside.

October came, and I launched the crowdfunding campaign on the third day. The swimming seasons would end on October 30th, Laura reminded me. I hoped for the best and kept checking the Windguru-website, looking for a window and hoping no more swimmers would show up. Meanwhile, I met one American writer working as an English teacher called Matt. He was from Utah and was working on his novel "An American in Cádiz." You can find his work online under the same title. He told me he had lived on a ranch ran by a Polish lady called Anya. The place was called La Leña, and the way he described it sounded like an enchanted community on the hills overlooking Tarifa. So, I went to La Leña to meet Anya, and Matt had been right. She cared for horses with love and had summer helpers run a small food garden.

We talked about toilets and she told me of her efforts, alongside British friend Janet. They ran an environmental organization which organized beach clean-ups and education campaigns in Gibraltar. They also lobbied both the Spanish government and European Union to curb air pollution in the region. Janet liked my swimming effort and asked her local Gibraltar TV Station to interview me.

Anya also told me about the wastewater treatment plant's effluent pipe five kilometers into the ocean in los Lances. I looked up online and saw YouTube videos of the works out in the ocean in front of Los Lances beach. We seem to ignore this, but wastewater treatment plants have an out pipe where treated water called effluent goes out. When the city is near the ocean like Tarifa, the cheapest solution is to lay a pipe in the bottom of the ocean a reasonably safe distance from the shore to keep the tides from bringing the effluent back on shore and pump it out into the ocean.

As I talked earlier, most plants in small cities only have secondary treatment. That is, the sewer will have settled in sedimentation ponds and then a chemical will be applied to kill bacteria and pathogens before this solution is discharged in the ocean. After this point, in both developed and developing nations, there is little follow up or way to know the full impacts of chlorine and other chemicals in marine life. Again, similar to Tulúm, tourists were contributing to the pollution of the very beaches they came to enjoy.

In mid-October, amongst the mounting tension, I did something I had been procrastinating for a while. I started meditating regularly. I had tried it with little results before, but I decided to give it an honest effort. I needed to organize my thoughts and control my emotions. From my triathlon races, I knew that my mental game was equally, if not more, important as my physical readiness. I had read once that swimming was active meditation, and I couldn't agree more. When you meditate, you must focus on your breath, and this is true of swimming. You must also empty your mind, and I've caught myself in this state while swimming.

It took years for me to realize what "being in the zone" felt like, even though I had experienced it before. This term is perhaps more commonly known as "runner's high" or "flow state," and it refers to an euphoria one usually experiences when doing cardiovascular exercises. I normally get in the zone at the end of practice. My mind is tamed, my body is wearied, and the swimming motions are in automatic. By then, my limbs temperature has equalized with that of the pool. My arms have pulled the water, entered and dragged it so much that I feel no barrier between myself and the water. My skin becomes liquid. My breaths are extended, and I come up only for a second to fill up my lungs, allowing me to go back under the surface. Then, the most striking thing happens, time doesn't exist. I feel as if I can swim forever. That's flow state and apparently also similar to being in the moment while meditating.

Most non-swimmers may think it is strange, but swimming is a strenuous sport. It takes a lot of strength to displace yourself through water which is 784 times denser than air. Open water swimming has been considered by many as one of the most challenging sports on the planet requiring mental and physical focus. The cold water is constantly draining heat from your body. Currents are also a factor, either helping or most likely, when you swim in the ocean, the waves are pulling your body in different directions while you're trying to swim in a straight line. One must always be aware of the surrounding both above and underwater. We must watch out for boats or jet skis while fighting off visions of sharks or monsters in our goggles' peripheral vision.

I was hoping to harness the benefits of meditation, so I started listening to guided meditations I found online. There was a 15 min-walking meditation that allowed me time to walk to Playa Chica, and once I got there, I meditated sitting and facing mount Yebel, the mountain on the Moroccan coastline. During my meditations, I did try to empty my mind, but when that did not work, I visualized the swim. I thought of myself in the middle of the Strait, swimming in the eternal blue. I imagined what my arms would look like and I would visualize myself looking at the palm of my hands and inspect my fingers against the deep blue ocean. I repeated this meditation every day. Another significant aspect of swimming is that my thoughts are freer when I'm in the water. It's as if everything is possible, *todo es posible*. I find myself calculating laps, my time spent swimming, how many kilometers or meters, then I convert the distances to miles. It's also easy to plan the future, think about stand-up routines I'd like to do one day, things I want to say to people. Swimming is like therapy to me. If I spend many days away from the water, I eventually get sad.

And when I do return to the water, the first feelings of being back in it are amazing. I love feeling buoyant, supported by trillions of tiny water particles. I like to stretch my arm as long as I can, elongating my body and thus moving more efficiently. It took me years to become aware of what each limb was doing and how to train my body to be streamlined, saving energy to swim longer. Swimming is about technique and style, not so much about physical strength. While training with *Team Hammerheads* in Florida, I was frustrated by having very skinny or chubby swimmers pass me when I felt I was stronger than them. I may have been stronger, but they were more efficient, and after one hour in the pool, efficiency will take you farther than power. When I get out of the water, after a workout, I'm at peace. My body is relaxed, and I am quiet. When I get out of the ocean, I am tamed, humbled and hungry. After eating, the first thing I want to do is take a nap.

However, I had to train and couldn't sleep much on October 18th. Alberto was taking my training very seriously, and he suggested we do a long swim along Los Lances, but to fit his classes schedule, I'd have to be in the water at 6 a.m. *Ouch*. Also, we needed additional help. We would need someone to drive the car from Los Lances to Ion Club at Valdevaqueros, the kiter's beach, where we planned to end the swim. It would be a 9km training session. Florian, the Swiss guy who lent me his suit offered to help by waking up at 5:30 and taking Alberto's car home, sleeping three more hours and then picking us and the kayak up at Valdevaqueros beach. Dankeschön Flor!

I woke up at 5 a.m., ate my breakfast: oatmeal with banana and water. Banana keeps muscle cramps away and milk ferments in your stomach when you're swimming so I eat cereal with water. Sometimes, I add orange juice to the oatmeal or cereal. If I'm craving sweets, I eat a dark chocolate bar. I love chocolate, by the way. At 5.30 a.m. I met Alberto and Florian at Los Lances, it was still dark, and I was humbled by their help. I helped Alberto drag the kayak to the beach and Florian drove back home to sleep. I got my earplugs in, swim cap on, goggles and zippered up my suit; I felt like a superhero again. Although there was a dim light on the horizon from the sun rising in the east, inside the water, light gets refracted, and I could see very well. I remember my first strokes distinctly. I was overwhelmed by gratitude. I was being helped by two guys I had met in the last three weeks and I felt blessed. They woke up early just to help me train for my swim, and swim I did. I felt strong and fearless. I saw bigger fish, but none came to say hi! We were in shallow waters for the first hours, and there was no movement in the ocean's bottom. I swam fast and consistent while Alberto stayed on my left side. I only stopped to eat two hours into the session at which point the saltwater starts burning the tongue. It's nice to have a sweet drink, but everything will taste salty after that. He gave me some cereal bars as I held on to his kayak, and we talked. I was happy. Alberto managed to get a good shot with the GoPro that Alex lent me and the gear from Christian. I used the images on my crowdfunding and Instagram account @whereisaldoj

I felt a current helping me. I had swum 9 km in 2:24 hrs. which would have put me in 4 kmph pace. That would have been too good to be true. However, it was a useful experience because it had been the longest time I spent in the suit and I needed to know if it fit me well or not. The chafing on my neck told me I needed to cut the suit so it wouldn't bother me during the swim. However, swimming assisted by a current didn't reassure me, so I asked Alberto for us to do a longer session. I wanted to swim at least 12 km, mind you that the Strait is 15 km long. From my triathlon training experience, I knew there was no use in training the entire length so close to the swim as I'd be risking muscle fatigue and injury. I was more concerned about keeping the 3 kmph pace.

So, we met again, a week later, this time at Playa Tarifa, near the spot where I swam the first time and got ejected from the sea. I was going to swim twelve kilometers to Valdevaqueros, and Florian didn't have to wake up at five am to pick up the car. But at the end of this session, the weather turned bad. The wind picked up considerably and huge waves were tossing me and Alberto's kayak away from each other. The water was turbid, and I couldn't see beyond my fingertips. We were almost done, and after swimming for four hours, I was exhausted. I didn't know if I had the strength to break the strong surf. The waves kept pushing us away from the beach. I had never been in such stormy waters before. I was reminded again of the danger of water sports and why kite surfers from around the world go to Tarifa specifically because of these strong winds. I kept calm and tried not to panic, after all, Alberto was in the water with me, or so I thought, before I realized I didn't know where he was. Then I spotted him far in the distance trying to kayak towards the beach. The wind was blowing salty water foam into his curly hair until it became straight. He must have been 200 meters away by now.

I was thankful for my years training open water in triathlons, my Red Cross Lifeguard training and I remained calm because I knew that if I panicked, then I could drown. The beach was about 100 meters ahead of me, but the waves were coming in so strong, and the current was taking me further downstream. I took a deep breath and started swimming, looking back to make sure no wave would crush me. I'd swim a little forward, then back again when the waves came. If a big wave comes and you think you will be hit, the wise thing to do is to swim under it. I learned this on my first beach holiday in Rio de Janeiro, Brazil. We called it "jacaré", alligator in Portuguese, which was the practice of ducking under incoming waves. I repeated this ducking a couple of times, swimming forward as much as I could, and then backward under big waves, until my toes touched the sand, and I could run out of the angry sea. I remember running ashore and collapsing on the sand. I was exhausted but glad to be alive. Someone walking their dog came to check on me, asking what I was doing in such waters. I had no energy to reply. I looked around for Alberto and found him dragging his kayak on the beach. I felt a huge relief and hoped that I wouldn't face a current this strong nearing Morocco.

On Friday, October 20th, the forecast was looking great for Saturday and Sunday crossings. Winds were going to be at three knots and waves shorter than one meter (3.3 feet) tall. I was hopeful this was going to be it. It was the last week of the month that I could swim across the strait. The three swimmers who had been given the window were from the UK: two men and one woman. They all had different paces which is not very good for group crossings because the faster ones have to wait for the slow swimmers. Their slowest person was swimming at two kmph so I thought I could swim with them. I went to Laura's office and begged her to ask them to consider me. She tried many times, but they did not want to swim with a stranger. I tried to convince them, sending my own messages, but they were unshakable. I was frustrated again.

Still, I was determined to train. On Saturday, Timo, a German nomad who had shared the apartment with Alex and I for a couple of weeks, wanted to stand up paddle board around Isla Paloma. I joined him, and we set out on one of my coolest practices in Tarifa. We met Manon and Ale, nomads from the Netherlands and Italy respectively, sunbathing at Playa Chica, and we got our gears ready. I got into my superhero swimming suit while Timo inflated his board. We swam clockwise around the island.

When we got to about 500 meters on this dotted line, as you can see, the ocean blue turns a darker shade. The sun was shining bright, and its rays reached deep, like white curtain lines; it was gorgeous. Different schools of fish would follow me while others swam past. I saw vibrant marine life on the corals near the island. Whenever I looked southwest, there was only blue. This is what swimming the Strait will look like, I told myself. My heart was pumping. I was so close to my goal, only five hours away, after two years of waiting.

When we reached the 1.25 km mark, we stopped to hydrate and fuel. At that side of the island, one can see the white painted lighthouse against the dark blue waters and blue sky. I wished more people had the opportunity to witness that sight as I did. I drank water and ate a chocolate bar. Timo also looked like he was having a lot of fun on the board, already sunburnt. But being in the water, below the surface, I felt as if we were in different worlds. After our snack break, we continued swimming and paddling around the island and then, as soon as I passed the southern tip, I faced a brutal current. This is the Spanish current that would take me to the Atlantic Ocean. Every day the Alboran sea is filled up in the morning, and in the evening the tide turns going back into the Atlantic Ocean. I was feeling that current towards the west. Tim floated very far from me, and since my eyes were at the water level, I could barely see him. I had to forget about him and focus on getting back on shore. I was spending way more calories than he was, and I had not brought additional food. I kept circling the island, looking at the jail on Isla Paloma which is said to house foreign migrants. In between my right-side breaths, I could see the barbed wire and boxy building and I could only ponder the irony of being imprisoned in this seeming paradise. I felt so lucky to be free, swimming around the Isla de Palomas, or Dove Island. I said a prayer for all the migrants who have died while crossing not only this strait but all the channels, borders and rivers like the Rio Grande between the United States and México.

8- ÁTTILA THE GREAT AND TITA THE PEACEFUL

Áttila sent me a WhatsApp message on October 23rd, 2017, introducing himself as "Áttila from Hungary", asking if I wanted to train with him. Laura had informed me about Áttila and that she shared my contact with him. She told me he was a pro swimmer completing the Seven Channels. That is, he had swum all the other channels in the circuit which I had read from Lynne Cox's book. Gibraltar was all he needed to complete his goal.

He texted me right before I was going to train in the Tarifa pool. Towards the end of October, I added pool sessions so I could swim longer and undistracted. I condensed the Gibraltar swim into an one-hour training session with buoys between my legs so I could focus on my arms. When you train, you will learn about training zones. They go from one to five, and they represent your effort level. Zone 1 refers to training at less than 50% effort. Zone 2 is 50% to 60% effort. Zone 3 workouts take place between 60% to 70% effort. Zone 4 is at 80% to 90% and, finally, Zone 5 is above 90% effort, also known as "all out." During the first 15-minutes, I focused on arms only at zone 4 by swimming with a buoy in between my legs. Then, I swam free-style for 30-minutes at zone 3 and the last 15 minutes was zone 4 with the last five minutes in zone 5. I enjoyed this particular training because halfway through it, I noticed a stronger swimmer on the lane next to me, and we ended up taciturnly racing the last laps.

When I got out of the pool, I answered Áttila's message, and we scheduled a swim for the next morning at Playa Chica. At the time, I had moved out of Alex's Airbnb, and I was housesitting Ellen's home up on the hills outside Tarifa and taking care of her dog Puka while she was out in the Netherlands. I wanted to meet Áttila because I thought that if I could swim with him, then perhaps I could also swim with him and since he had a se-cured swimming window, I would have more chances of swimming. Swimming in a team would also half my crossing fee of two thousand euros. The fundraising was going well but I was still short of the goal to cover both the toilets and the swim. All the while, the clock was ticking, and we had only one week to attempt the crossing.

Remember that Ellen's house was almost a "Camino de San-tiago" away from Tarifa? I had to cycle the eight kilometers and a half into town, and I was running late to meet Áttila. First im-pressions last forever, I worried. I picked up my borrowed bike and raced it down the hills yelling and screaming out of excite-ment, carelessness and joy. As I sped down, I didn't have time to remember that Paulo Coelho's famous shepherd boy Santiago from "The Alchemist" actually starts his journey in Tarifa. I had only read the first three pages of the book but I'm sure that I had finished it, I would have found more similarities between us. Wasn't alchemy concerned with transforming things into gold? Well, as per Carol's book, urine is liquid gold and feces can be composted into humus which has been called by many as black gold. Unbeknownst to me, similarly to Santiago, I was riding my own journey.

When I arrived at Playa Chica, I saw someone swimming back

and forth in between the two yellow buoys where I started my swims, and I assumed it was Áttila. I jumped in and swam towards the head bobbing in the water. We greeted and I suggested that we swim to Isla Paloma, on my usual 650 meters course and he asked me: "Is it safe?" to which I said yes because by then I knew every rock on the ocean floor on the way there, and there was no time to explain the swimming dog incident.

He was built and had a strong pull which quickly left me in the back, and I lost visual contact. When we got to the island, I swam to meet him, and we swam back. Again, he was incredibly fast. There was no way I'd swim with him, I thought, as I swam towards the shore. I saw him standing, and I swam towards him, and when I got up, I noticed he was only 1,55cm (5 feet) tall. In the water your height doesn't matter, and I guess neither does it outside. Especially in long-distance swimming propulsion comes from the arms, not the legs. Perhaps during a sprint race, one can benefit from extra kicking but not for kilometers.

Being short is not a disadvantage. Áttila lifted weights, and he could move effortlessly through the water. I could only keep up with him at his comfortable pace for a couple of minutes, before reality kicked in, and I was left in his bubble trail, eventually so far behind I could no longer see him. It was not encouraging for me. My goal, though, was to swim at 3kmph, which would take me across the Strait in five hours. After the first training session, we went for coffee at Café Solsticio. He had a VW Golf rental and did not trust the sketchy car keepers at the illegal car lot next to the beach. I asked him many questions, and at first, he seemed cynical, but it became apparent it was his Hungarian personality.

In England, I had worked with two Hungarians and it prepared me to understand Áttila. I asked him why he was swimming and what kept him going, he gave me direct answers, which at first seemed cynical. Swimming was his job, his trade, what he did, and he did it very well. He was sponsored by big companies, both in Hungary and abroad. He gave motivational speeches, was fully booked with media obligations and had a wife and two kids. Swimming was his life. As our training sessions piled up, I came to know him better. He shared with me all the sacrifices, such as the time he slept in a post office before his swim meet and the precarious conditions of his early days.

Now, 2017, it was time to enjoy the fruits of his labor, fair play as my British friends would say. He was staying at a nice hotel in Algeciras, a 30-min drive from Tarifa because he wanted to enjoy the comforts of a bigger town. He would always say, swimming for him was like a "business". He invested in his career, and now it was time to reap the dividends. He needed to stay on top and manage it well to keep it rolling. As the training sessions went by, I got to see a more relaxed Áttila with an infectious laughter, sometimes mischievously so. I called him the "Hungarian Torpedo", and he nicknamed me "10 Euros" because that's how much my room cost per day. We both drew fun at the opposites in which we found ourselves, yet we were united by the water. He was a sponsored athlete leisurely swimming across Gibraltar, while I was holding on for dear life and struggling to make this swim possible, both financially and physically. However, when we were swimming, it was all fun. He was also uncertain about his swim because of the weather, but we still trained and hoped everything would work out in the end. I focused on my time and my pace but kept training with Áttila to learn from his experience.

The we were joined by new swimmer Zach. He was from the U.S. and was also hoping to swim in the last days of the season. Unlike them, however, I didn't have a window in the 2017 season, remember? I was the intruder, desperately hoping for a day to swim. Both of them had been given a window to swim. Having another swimmer join us could not be good news for me. The three of us trained together, and Zach was also clearly faster than me, able to hold a 3,5 kmph pace or faster. Áttila was swimming at 4,1kmph, and I was barely keeping my 3kmph. There was no way I'd be swimming with either of them, or both of them. I was too slow. I wouldn't want to tarnish Áttila's record with a slow swim or frustrate either of them by making them wait for me. I was going to have to swim alone, which meant I'd have to pay the full fee for the boats €1,905.00, but it also meant that I was at the mercy of the weather gods which hadn't been too nice so far that year. The more I thought about the swim, the tenser I got. Every time I trained, I tried not to think about the likelihood of having wasted my time and money in Spain, when the most probable outcome was having to come back the following year, like the other hundreds of swimmers before me. Still, I had faith, somehow, someway everything would work out.

My friends in Tarifa, although well-meaning, only added to the pressure. Every time I saw each of them, they would ask me: "So, Aldo...when are you swimming?" My answer was either "Next window," or "I don't know." The fact that I wasn't a kite surfer like most of them only added to my singularity in town. I was becoming a joke, the guy who was never going to swim to Morocco. Sometimes, I like to scream in the water when I'm swimming. Normally it's because I'm happy but those days I screamed out of frustration, making many angry bubbles and almost deafening myself. My training wasn't going very well. I was training up to an hour per day to avoid straining my right shoulder which was giving early signs of fatigue. I was feeling an ominously familiar pain that reminded of me my last swimming hours in the swim around Key West.

On October 25th, I gave an interview to the Gibraltar TV. The news reporter asked me about compost toilets in México and then the obvious question: "When are you swimming?" Looking back, I don't know where I got the faith and courage to tell him that I was hopeful we would be swimming in the next days. The interview should be on YouTube. I came back to Tarifa from Gibraltar and was excited to tell Áttila, but he wasn't impressed; he's done too many interviews. We went swimming at Playa Chica for 45 minutes the next morning in what I felt was cold water. He told me not say the word "cold," instead he said the water was "fresh." It's a habit I'm adapting for myself as part of the mental game. If I say that the water is cold, then I will feel cold.

We were regularly checking the wind forecasts and trying to guess who would swim and when. With five days till the end of the season, it was safe to assume that no more swimmers would show up. I guessed Áttila would swim first, then Zach and hopefully, me. But this is what the wind forecast looked like on the 26th of October:

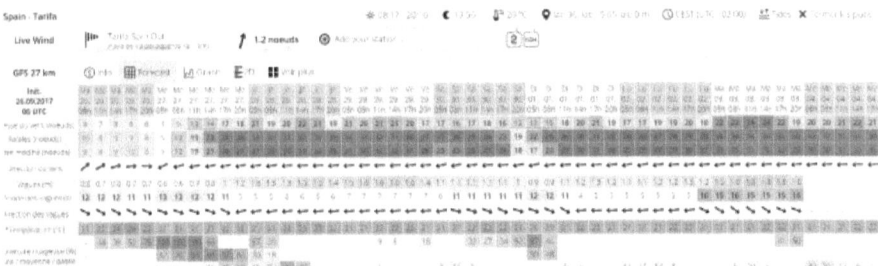

It didn't look like anybody was going to cross the Strait anytime soon, so we went training. This time we chose Los Lances, in front of Café Surla. Looking out from the beach I remembered the footages from YouTube. The wastewater plant effluent pipe was somewhere out beneath the surface in front of us. We did a different training session this time. There were some buoys in this part of the beach, and we figured they were about 500 meters apart. We swam with the current going one direction and then came back against it. We did this set three times for a total of 52 minutes in the water.

The first lap was okay, I took six minutes to swim one 500 meters lap at a relaxed pace and then twelve minutes to swim back against the current. This was the warm-up. On the second lap, I swam faster at four minutes to get to the second buoy, then 13 minutes to swim back the 500 meters again against the current. I tasted my breakfast for the second time while trying to keep up with Áttila at his easy pace. On the last lap, I swam the first leg at four minutes and went for consistency keeping up with the previous pace of 13 minutes. Then I swam all out from the buoy to the beach for another 150 meters. It's always good to end the training at a sprint pace, especially for triathlons, but it was also a reminder that I'd have to reach very deep in my mental focus to prepare myself for the last kilometers reaching Morocco and breaking in the current along the coast. "It's all in my head", I told myself.

On the 29th at evening practice, I heard that Zach and Áttila trained with a woman who had swum from coastal Spain to Ibiza, the distance of 90 kilometers (56 miles), in 28 hours. I was utterly impressed with such a feat and wanted to swim with her. Next morning, Áttila and I found her patiently waiting with her equally gracious partner Francisco. The first impression I got from meeting Tita was of complete peacefulness. I felt as if all the ocean's tranquility had been transferred to her during the countless hours she spent in the water. She was ready to go, with cap and goggles on her forehead, peacefully waiting for us with warm smiles and hugs. I rushed to clumsily put on my two caps and oversized goggles while running into the fresh waters of Playa Chica. During the first meters, Tita flanked my right and Áttila my left shoulder, and I felt incredibly out of place, swimming with two legends of open water swimming. If you added how many kilometers these two must have trained, I'm sure they could have wrapped the globe a couple of times.

I also felt a newfound admiration for open water swimming for such honor of sharing the water with these two giants. All this thinking took place in seconds because I was trying very hard to keep up with them. The first 50 meters went by, and I was nervously excited, stretching very long, trying to streamline as much as I could to swim efficiently, but I also had to swim very fast to keep up with them. The first 100 meters went by, and I started dragging. I kicked hard and recovered some ground. I found myself squeezed in between them and the more I thought about it, the more out of place I felt. I started dragging behind them, until eventually being left in their bubbles and foam created by their kicks. I was too slow. I rebuked myself, just swim; don't think! They waited for me on the other side at Isla Paloma, chatting while looking at me. Somehow, I had swum to the right and ended near the rocks. Áttila whistled, allowing me to find their heads bobbing up and down in the water. I ashamedly swam towards them where they joked about me not being able to see despite my scuba goggles. They laughed, and I smile awkwardly.

It was time to swim back the 325 meters. I hurried to get an early start, and they followed. This time, Áttila was on my right and Tita on my left. Again, they were swimming at an easy pace, and I was in awe of this moment where I could literally rub shoulders with world-class swimmers. The waves knocked us against each other sometimes. For a couple of strokes, Áttila's left arm and mine synchronized, and we kept up at the same speed. I wondered how long this would last. I tried not to think about the incredible honor and sensation of this moment: I'm swimming next to Áttila and Tita Lorens. Don't think about it. Swim! Swim! Swim! But eventually, I was left in their bubbles and, again, they had to wait for me. Tita asked me how many hours I've trained, and I exaggerated a bit by telling her one to two hours every day. She looked a bit concerned at me. Meanwhile, Áttila started recording a video in Hungarian for his followers.

Bianca invited me for another horseback ride in the afternoon which I eagerly accepted. This time I rode Evita, the other white Spanish war horse and we were accompanied by her son Único, and Guapa, her German shepherd. As soon as we reached the beach, the sun started setting, the six of us admiring nature being painted in front of us. It was like a dream. Couples and photographers were trying to capture the beauty, while I was feeling it by closing my eyes and inhaling deeply. Kite surfers, still dripping saltwater, stopped folding their sandy kites to appreciate that moment. I was living in a dream.

9 – TENSION

I t was November 1ˢᵗ, and the swimming season was technically over. I packed my bags, but I was not leaving Tarifa,
yet; I was moving out of Ellen's place and into an apartment in town where I could be closer to Playa Chica and also
be able to swim in a moment's notice. The room only cost ten
euros per day. The place was within ten minutes of the dock by
foot. It was a really nice find through Vince, an Italian guy who
was working at a pizza place in the city center. This is what the
weather forecast looked like for the next days. A lot of wind for
the rest of the week and a wind break on Saturday. I wouldn't be
able to swim in this scenario.

Áttila was the strongest swimmer, so he was comfortable with
swimming in strong winds. Then Zach could swim after him
and then, hopefully, me. We were all waiting to hear from Laura
and her experienced captains who had been taking swimmers
across the Strait for more than a decade. We trusted their judge-
ment and experience.

All we could do was pray to the weather gods and train. Áttila
posted a message to his followers on Facebook, notifying them
of his swim the next day. I couldn't help but feel jealous and

wish I could be doing the same. I used the translator feature of the post and was surprised to read he would be swimming with Zach. I hadn't expected them to be swimming together. Áttila was clearly stronger than Zach and Thought he wouldn't want his swim ruined or his time slowed by another swimmer. Nonetheless this was great news for me—or not. The fact they decided to swim together made it more likely for me to swim on the next possible day, or it could have meant they were afraid the weather would be changing so drastically that they wanted to attempt a riskier and earlier swim.

I didn't want to think about it. At this point, I was emotionally distressed. Thankfully, there were digital nomads keen on having fun in Tarifa. There was a Salsa lesson that night and I went out with my friends and tried to relax and forget swimming for an hour. Then, I went home and something interesting happened. I fell asleep and dreamt that I was buying Powerade and the foods which I usually eat during my swims, like banana and chocolate. I thought it was a strange dream because I don't like drink Powerade.

Next morning, on November 2nd morning, I went to the Tarifa library, to use their Wi-Fi and continue my fundraising efforts. The *mirador* was nearby and I went there first to get a visual of Áttila and Zach's progress in the water. I squinted my eyes, yet I could barely see the inflatable and the white dot that must have been the boat Columba. I assumed it was them in the middle of the Strait. Cargo ships seemed like giants next to them. Then, I logged in the library Wi-Fi on my phone and opened WhatsApp. I was getting a lot of messages from all my friends and fundraising donors, so it took me a while to see this one from Laura:

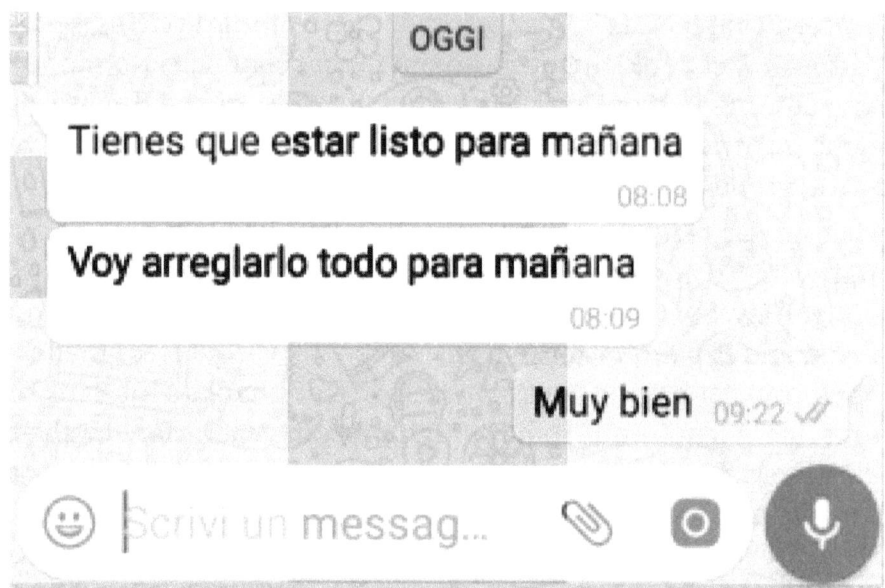

I couldn't believe my eyes! She was telling me to get ready for tomorrow, Nov 3rd. Everything would be ready for me to swim the next day. I was ecstatic! I went on Windguru and saw this:

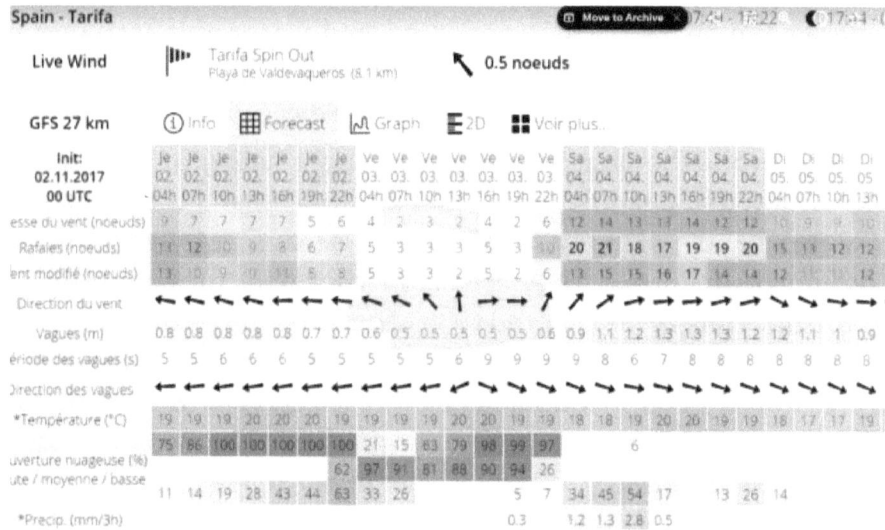

Wind speed of three knots on Friday! That was as if the sea had opened for me to walk through in Biblical terms. My jaw dropped. That wind forecast dropped from 13 to 3 knots overnight. Meanwhile, my one-man social media campaign was giving results and I was being featured in swimming articles, sport pages on Facebook and other social media channels. No money was coming in from strangers, though. 99% of the donations came from people who knew me personally. I was very thankful to my friends.

I had to get ready now just like in my dream. I left the library and went to the ATM to withdraw the money to pay Laura, but it didn't work. I had received payment for my translation work, and it paid for more than half of the €1905 fee. I made a bank transfer paying huge transaction fees, thanks a lot Bank of America. Then I went to have high protein vegetarian lunch at Chilimosa, the local veggie place in Tarifa because I didn't want to have any meats churning in my stomach during the swim. Meat proteins take considerably longer to digest. I also had to get a lot of last-minute preparations. I bought banana and chocolate like in my dream, talk about Déjá Vu! I made a mix consisting of peanut butter, dark chocolate and banana that I was going to be slurping during the crossing. The sweets would give me energy and abate the saltiness. I packed water and Áttila gave me an energy drink powder from his sponsors. I was breaking the rule not to try anything new on race day.

Then, I broke the same rule again. I bought new gear. I was very familiar with Speedo's Vanquisher 2.0 goggles and I loved them. I wasn't happy with my scuba goggles because they were leaky and foggy. In Europe, they are very expensive compared to the U.S. prices and I only wanted to get them for the swim. I also got ear plugs because the fresh water really bothers my ears. When I went to try the speedo goggles, I realized they were not the same model to which I was accustomed, but I had no other option. I'd be swimming with beginner level goggles.

In the afternoon, Laura came back from crossing with Áttila and Zach, and we met at her office. The same office I had been sitting with Marianne seven weeks earlier. But Laura and I were both in different spirits. She explained to me how the crossing would be. I'd arrive at eight a.m. at the Tarifa dock, get dressed and jump in the water by eight thirty. There would be two boats following me. One Zodiac with my food that would be on my right side, and a larger white boat named Columba in front of me. I'd have to follow the white boat because it would be pointing the direction. I would set the pace. "Okay fine", I said and transferred all the money available in my account to hers.

Then I went on Gofundme and shared my last update with my followers. I gave them instructions of how to follow the Boat GPS tracking system online and thanked them for their financial support. I felt so lucky to be able to swim for toilets. I cycled to Playa Chica and got in the water with the new goggles, earplugs and my superhero swimsuit. I needed to test the new gear. It was getting dark again, but I didn't go very far. I wanted to make sure the goggles weren't leaking. They passed the quick test and I went to get dinner, again, at Chilimosa. Next, I went for the doctor checkup. Since I had not confirmation, I hadn't gone to get a medical clearance until now, 14 hours before the swim. I walked into the Spanish doctor's office when I should have been going to bed. I'm not sure there was anything this doctor could have said that would have stopped my swim, but every time he went out of the office, I was nervous. He was just talking to his assistant who was also his wife. He asked me to do some old school and weird exercises squatting, holding my breath and connected me to an ancient computer with thick cables.

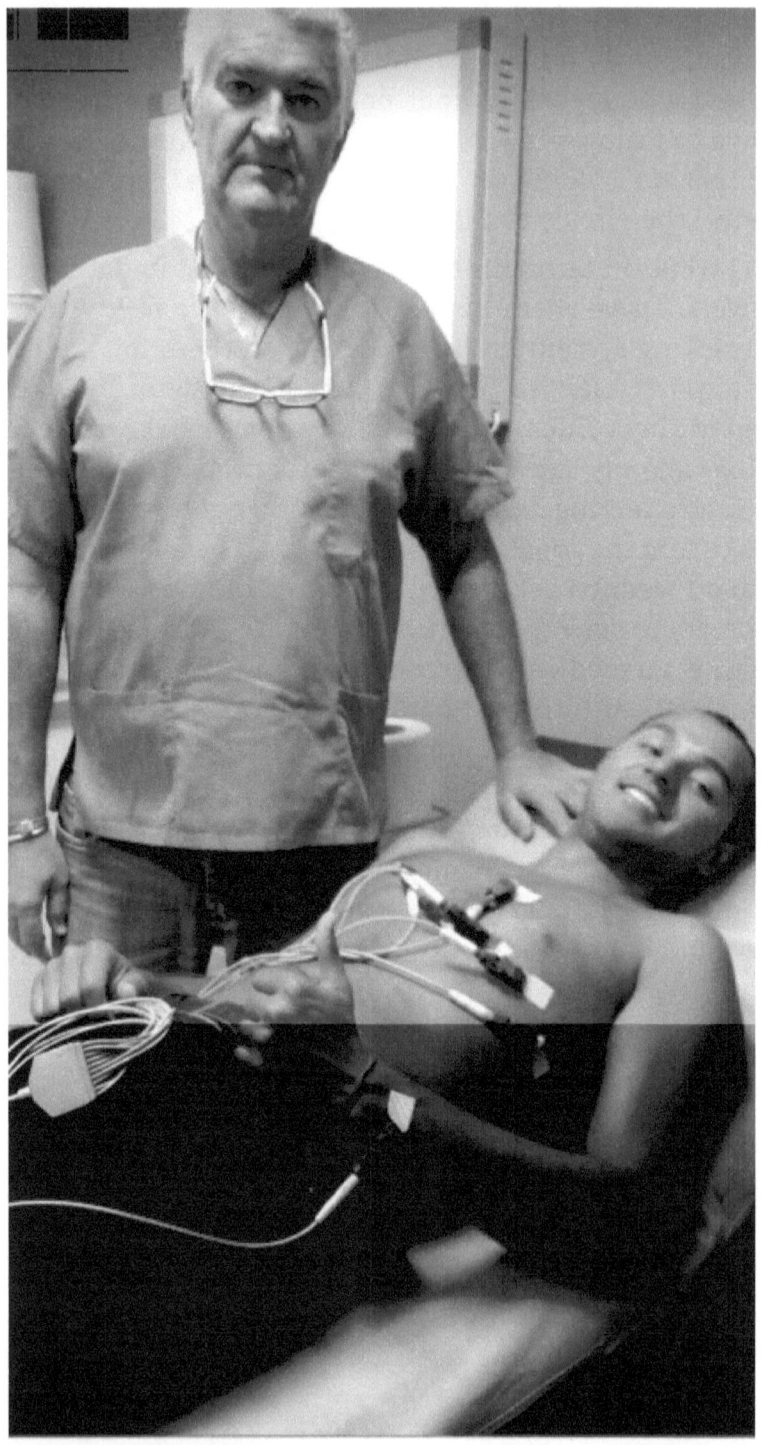

I breathed deeply like in my meditation and tried not to think. Then he came back and said it was all good. "Muy bien!" I said, grabbed a paper slip, paid the fee, and walked back home which was only one block away. My living room table was full of bananas, honey, chocolate, peanut butter and swimming gear, all the stuff I had purchased during the day. I prepared my gooey mix, hydration bottles, set my alarm and went to bed.

10 – SWIM

I was scheduled to be on the deck at 08:00 with the swim starting at 08:30. From my triathlons and swimming experience, I had learned to eat breakfast at least two hours before going in the water. I was up at 6:13 and had granola cereal with chocolate, honey and banana. I picked up my gooey nutrition bag, hydration bottles, packed them in my bicycle and headed to the port. On the way, I stopped at Baraonda to steal their Wi-Fi and send a last message to my friends and see if Matt was up as he told me he would like to meet me at the dock.

Surprisingly, I wasn't nervous, and I was happy to see Matt walking up on deck. I started talking with him trying to distract myself from what I was about to do while changing into my superhero suit. He wished me well and we hugged before Laura started giving me the final instructions. She introduced me to her experienced team whom her father had trained over countless crossings. First, I met Capitán António who would be piloting Columba, the larger white boat I should follow. Then I shook hands with Fernando who would be piloting a black inflatable, Laura would be with him. We left the port and headed to Isla Paloma, the starting point of the crossing as I had been informed by Laura two months ago. I remembered not having payed full attention to this part as I wasn't sure this day would actually come. I started stretching and taking pictures and videos of my last dry minutes.

I felt the boat rocking as it left the dock. On the left, or portside, the sun was rising behind a statue of some patron saint that sits at the end of the dock walls. The sky was glowing orange. Over the liquid horizon, I saw the water surface mirroring the heavens and mount Yebel completed the picture frame. I was about to swim over to Africa. At starboard, or my right, Isla de Paloma grew into focus and we fatefully drifted towards it.

I inserted the earplugs in my ears, muffling all exterior sounds while amplifying the sensation caused by my own breath. I inhaled deeply and stretched the first latex swimming cap as to fit my head. Then I carefully placed my swimming goggles, making sure the two straps were far apart from one another and in the appropriate place in the back of my head, as I've done a million times by now. Next, I opened my mouth and massage my jaws, an old habit of mine. Finally, I put on the second swimming cap to keep my goggles and first cap tight in place.

We arrived at Isla Paloma and Capitán António turned the boat around as he must have done countless times. This was at the same spot from that training session with Timo where I saw the deep blue of the Strait. It had paid off to train the week before. Instead of being angry, I had prepared myself for this moment. I felt ready because it felt as if I had created each second that I was living, like in a lucid dream. Then, Capitán Antonio raised his right arm and index finger pointing to a seagull precariously perched on a rocky edge of the island. His mouth moved in slow motion and unintelligible sounds came from it, but their meaning didn't matter because I knew exactly what I had to do, touch the rock where the seagull was sitting and swim from Spain to Morocco.

I jumped overboard like a cannonball. I went deep and when the bubbles dissipated, I saw many fish of different tones of blue. They must have been as startled and confused as I was because they stared back at me with their wide, unblinking eyes. I said: "Hello fishy!" and swam towards Isla Paloma unintentionally scaring away the seagull. I touched the mossy rock and heard a muffled whistle blowing from the inflatable. It was finally my time to start swimming. I took my first arm stroke and got my head in the water where I could see that world of fish again, all around me. When I came back for air for the first time, I saw Fernando starting his chronometer and Laura taking a picture of me.

I couldn't fully appreciate this moment's beauty because this was the most nervous, I had been, up until then. I felt as if my heart was in my throat. Laura and Fernando were to my right in a zodiac christened *Duende del Mar* or sea elves. Remember Oaxaca's long-distance runner Fernando who had hallucinated with other types of *duendes*, forest elves? I was not afraid as this elf from the sea was actually going to help me. I felt safe and started swimming towards Columba, pointing the direction in which I should swim.

At this point, I wished that each one of my friends and supporters could have been in the water with me. I felt brave precisely because I was staring at my fear. My heart was pumping hard. I then remembered to inhale and make big bubbles. That familiar vibration in my head calmed me, and I felt that everyone in the world needed to experience jumping in the ocean at 8:30a.m. and swim to Morocco. I tried not think about what I was doing and told myself: Swim.

Laura had stressed the importance of swimming fast for the first two hours in order to break away from the current hugging the Spanish coastline. She said I needed to swim fast and strong and avoid unnecessary pauses. If I had to stop for feeding, then it would have to be less than one minute in order to not be taken off course by the current. At the same time, she reminded me to save energy for the end, echoing the advice I had heard from experienced swimmers.

I was approaching the first hour mark and I wasn't hungry, in fact my breakfast kept coming back up. So, I told Fernando and Laura that I wasn't going to stop to fuel by moving my index finger to the left and right. I saw one piece of plastic wrapping drafting within reach of me and I swam a bit to the left to pick it up and stick it in my wetsuit sleeve. One hour and a half in and I was still feeling full, which gave me the confidence to swim thirty more minutes before stopping to feed at the two-hour mark. I avoided looking at my watch as it makes time flow faster.

I originally had planned for Marianne to feed me but since it took me two months longer to swim, she was now in France, and Laura volunteered to do it. She handed me a Ziplock bag with my banana, honey, chocolate and peanut butter gooey mix and I washed it down with the hydration solution that Áttila gifted me. This feeding took less than one minute, and I went back to swimming.

Since I hoped to complete the swim within five hours, my midway point would be at 2:30 hrs. and I started periodically looking up to Fernando for his sign that I had reached the halfway point, something he told me he would do. Then, it occurred to me there was a small chance that I could die while swimming. There's always chance for something to go wrong and although we had taken the necessary precautions, we would still be crossing the Strait of Gibraltar. Yet, this realization didn't make me nervous. On the contrary, it was It was a sobering thought, not because it was dangerous but because at that point, I realized I had found something worth dying for. The newspaper headlines would have been funny: "Man dies swimming the Strait of Gibraltar for toilets in México." I guess there would have been many questions from puzzled readers. I would never have been found swimming for water flushing toilets nor water-saving showers though, and that realization impacted me. I had found something that was guiding my life, something that gave it meaning and purpose.

I felt rolling waves helping me, and in between the ups and downs I could spot a cargo ship or another boat in the distance. All the while, I was following Columba's direction and making sure I was not too far from *El Duende*. 2:30 hrs. came and went, and Fernando didn't make the half-way sign, then 2:35, 2:40 and I was hoping he had forgotten to tell me. Then at 2:50 hrs. Fernando made a "T" sign by placing his left arm horizontally and jabbering it with his right arm, holding it perpendicular. I realized I was 20 minutes late. It was not good, I thought and calculated that if I kept this pace, then I would be in Morocco at 5:40 hours and this was also not so good. This pace would take me across in almost six hours, below the 3kmph rhythm asked by Laura.

For the last 3 days I had felt that slight and familiar discomfort in my right shoulder, so I was being careful not to tear any ligament. I must have compensated in my kick because I was feeling soreness in my kneecaps. Meanwhile the wetsuit was chafing my neck on both sides. It felt as if it was being serrated with a plastic, dull knife. I felt the discomfort, embraced it and allowed it to push me to swim faster so the pain would end. Although worried about my pace, I knew that after having passed the halfway point, there was less of a chance of being pulled out of the water for being slow.

I had more than five hours remaining to safely conclude the swim as per the associations eight-hour cut-off limit. I reasoned just 2 hours and 40 minutes more and I would be done. I then told myself to enjoy each minute of the swim. This was what I had dreamt, trained and believed in. Then I witnessed something I had never seen before. A tiny green light drifted below me. It was giving off light, so I didn't think it was a plankton. It looked like a green shooting start underwater, against the black deep ocean. Whatever it was, I felt the need to make a wish and smiled underwater as I have done many times before. Three hours passed by and I stopped to hydrate only, no food. I took less than 30 seconds and got back to swimming. I remembered my mantras from Yoda, and Ayrton Senna

Yoda said: "Your focus becomes your reality," and I focused on swimming ignoring the pains. And I like to remember a famous quote by Ayrton Senna, the race car driver where he says: "*Sempre faça tudo com muito amor e com muita fé em Deus, que um dia você chega lá, de alguma maneira você chega lá*" which means to go about your life with a lot of love and faith in everything that you do because one day you will get there, somehow you will get there. I was living those words with my strokes and getting energy from them. I realized at that moment that my faith had made all of this possible. There had been no need to worry during my training, doubting myself, fearing I wouldn't be able to swim. I remembered Ellen's advice to live out of trust, not fear. My job was only to train and I was proud of myself for having done that. During my two months in Tarifa, all I needed to do was to believe that I was going to swim, and here I was swimming with a smile on my face underwater.

Four hours swimming passed, and I stopped to fuel and hydrate which only took 55 seconds. I told myself: "I will be able to relax in one more hour." Not because I would be done but because I only had two more hours left and the very last hour would be the best one. I managed my shoulder and neck discomfort which is athlete's speak for repressing the pain. Then, I stopped feeling my neck at all. It was a cloudy day and I was thankful because being out in the sun for so many hours would have tired me faster. I decided to swim backstroke for a while in order to relax the muscle groups associated with freestyle. It was important to change things up a bit, look up at the sky, swim on my back and look at my toes kicking water up in the air. I spotted another cargo ship in the distance. I could see Morocco already, Yebel was behind the clouds. By now, the sea was not so calm anymore; I bobbled up and down with the waves, sometimes I couldn't see either boats supporting me. Both of my knees were in pain. I took long strokes, stretching my arms as long as I could.

Five hours passed and I thought, this was it, the hour that every-one told me about. Let's break the Moroccan current. I hydrated at the top of the hour as I didn't want the food weighing down on me. Fernando then told me that I had two kilometers to go by raising his index and middle finger, making a "V" of vic-tory. I calculated: 2 km at 18 min/km and I should be there at 5:36. When you have a long swim, it's fun to calculate the time remaining and speed. It helps to keep the brain entertained. After 20 minutes I looked at Fernando hoping for him to raise his index finger, meaning there was only 1 km left to go but he didn't look away from the Moroccan coast. Before entering the water, we had agreed that he would let me know when we were near the Moroccan rocks. I continued swimming, right side arm stroke, filling up my lungs, and looking forward underwater.

Columba and *El Duende* were frantically navigating around me and I kept calm despite the mounting tension. I looked to the left and I saw a small island outcropping and I recognize it from the pictures of Áttila's successful swim the previous day. The island looked very far though, but it was to the left, while the boats were to my right. Capitán António precisely told me that at this point I would see land but that I must trust them and follow the boats. This conflict between following the boat or going to the shore which looked closer was what made the last kilometers mentally challenging. Five hours and 30 minutes passed, and Fernando gave me the one-kilometer sign. I gave him a mental "OK" by breathing to my right, and pulling water, staring at the bottom of the ocean which I could now see.

This last kilometer was the hardest, as I had been told. I had to gather all the strength I had left and focus on my stroke. Following the boat while the island seemed to near me very slowly. 5:40 hours passed and I stopped looking at my watch. I also stopped thinking as thoughts demanded too much energy and my power-hungry brain was shutting down. On the ocean floor's, I saw the seaweed being blown back similar to the palm trees being blown back on the beach before my swim around Key West. I realized I was swimming against the current. One tiny fish below joined me in solidarity, whipping his body to keep up against the water pushing both of us back. I was approaching the island, painfully slowly against the strong current. I kept swimming and inhaling deeply, making bubbles and feeling my head vibrating and at times closing my eyes for long seconds.

I don't remember much of what happened during this time because I was so exhausted and cold, I felt that my brain was almost off. I could no longer feel my fingertips. I was not counting strokes nor looking for marine life. I was beyond flow state, just swimming and breathing, living. Then, I heard a faint clapping from *El Duende*. The clapping got louder, and I realized it was Fernando letting me know I had reached Morocco and when he caught my eyes, he pointed me to the rocky island that would finish the swim. It got shallow suddenly and I got up and walked on the African continent. I was surprised I had made it to Morocco! I climbed up a bit and looked back at them in disbelief. It was over.

But it wasn't fully over yet. I got off the rock and the approximate 30 meters from me to Columba felt much longer. It felt like the longest warm down given to me by any coach I've had. I was so exhausted I couldn't swim back to Columba. I treaded water, paddled like a dog and rested on my back. Finally, I reached the boat and they helped me up and headed Columba towards Spain. I didn't have the strength to hug, nor high-five anyone. I couldn't even take off my cold swimsuit. All I could do was to remove the top part of my suit and wrap myself with my traveler's towel and put on some warm clothes. I was shivering badly, showing early signs of hypothermia. I wouldn't have lasted much longer in the fresh water.

As an athlete you don't want to under-train risking losing the race. On the other hand, over-training can lead to injuries and is a waste of time. Therefore, it's hard to find the right equilibrium that will get you to your goal. Sometimes I felt lazy because I wasn't training long hours like my experienced friends suggested. However, looking back I realized my training had been efficient. My training had been perfect for this distance and time in the water. My five-hour goal was set by and there would be no harm in finishing one hour later. It was still well within the limits for the crossing. Ultimately, I was happy I had left it all in the sea; my coaches would have been proud.

When the boat docked back in Tarifa, Laura came and congratulated me, and then hugged Antonio and Fernando celebrating the closing of the 2017 swimming season. Heavy clouds were rolling in and hours later, a rainstorm fell over Tarifa which signified the first storms of winter, the natural closing of the crossing season. If Zach had not swam with Áttila the previous day, I wouldn't be swimming on Saturday, even if the weather permitted. I was truly blessed; I had been the last person to swim in 2017.

11 – BUILDING THE TOILET OF THE FUTURE

I flew to México City, Oaxaca and then took a bus to the Isthmus of Ixtepec's bus station. This was the region that suffered the most from the earthquake in the state of Oaxaca. I was welcomed by Miguel, Heifer's man on the ground. Heifer was the nonprofit with whom I had built the first toilet in Oaxaca the year before. He drove me to his house in a white truck. We parked the car inside the garage and walked up the stairs avoiding his black and very angry dog barking at me. His girlfriend wasn't home, and the house looked like a single bachelor had been throwing parties for weeks. But the true reason for the mess emerged; there were major hairline cracks on every wall. The picture frames were placed on safer places such as the tables or ground.

In México, normally the women cook while the man waits, but since his girlfriend was coming back from out of town the next day, he quickly put together a meal by re-heating tortillas and *frijoles negros* (black beans) with eggs. He asked me to get a jar of freshly squeezed mango juice from the fridge and set it on the table. It looked like liquid gold, viscous, creamy, and the mine was the mango tree in the patio guarded by the dog. The tree was loaded with ripe mangoes which exploded as they hit the floor or his dog's tin roof shed. Mexicans take their tortillas very seriously. They're made from Corn flour, or *Harina de Maíz*, and constitute an integral part of the Mexican diet and culture. Before every meal, they ritually warm and slip them inside kitchen towels converted into envelopes to conserve their heat. I've never been offered cold tortillas and I assume it would be offensive to do so. Miguel handed me his home-made salsa, spicy sauce made with local chili peppers, after all who eats tacos without *salsa*? I gratefully ate this banquet with my copy of the Humanure handbook placed next to me. I was going to tell Miguel about it, but I was so tired from the bus ride that I could no longer speak Spanish. Speaking a foreign language requires a lot of energy and I've found that as the day progresses and I get tired, I slowly lose my linguistic abilities.

The next day, I woke up rested and ready to speak Spanish with Miguel's mother who showed up and made the whole house smell like something beautiful was cooking in the kitchen. It was a love-filled breakfast even though it contained the same but fresher ingredients from dinner: tortillas, eggs, *frijoles* and more mango juice. I finished breakfast, said *Gracias* to Miguel's mother, witnessed her blessing her son like my grandmother did with my father. We went down to the mango tree patio where his ferocious dog lived. Again, I stayed close to the wall, avoiding his dog jumping and barking at us. From the mango-splashed patio, we walked to the garage gate which he had to unlock so we could access the car. As I watched him fumbling for his keys, I appreciated the building code and fire safety regulations in the United States. in most cases, one should always be able to exit a building without keys precisely because of fires and earthquakes. As I'm comparing building codes and enforcement, Miguel reveals that most people died not during but after the earthquake and its aftershocks because they couldn't get out of their trembling houses. I silently prayed for no aftershock during those precious seconds he couldn't unlock the gate.

We got in the truck and drove out through the rubbled city. I noticed most houses were damaged or condemned. Mountains of debris were placed at the end of every road for collection by the dump trucks. It looked like a warzone but the military cars patrolling the streets reassured the nervous among us against possible looters and lawlessness. I was informed there had been more than 2,000 aftershocks since the earthquake in September but pedestrians on the streets were trying to go about their daily lives, walking around the wreckage and queueing up at the hardware store to repair what they could.

To my dismay, as we were touring the city, Miguel told me there was not much interest in the compost toilets. He said only six families showed interest. I was very disappointed at first but then I was reminded of this same feeling when I was visited an informal settlement in the island of Cozumel, off the coast of Playa del Carmen. This island is famous for its Caribbean beaches and international resorts. There is a Half-Ironman there every year. I was visiting the island in company of friends I met in Cancún after my TEDx and I wanted to see the sanitary conditions in the poorest neighborhoods. I approached an elderly lady to ask her if she had toilets to which she said yes. I asked if she had any type of wastewater treatment or containment to which she said no. Instead, she expressed pride in her own water flushing toilet.

She admitted to be stealing electricity from the grid with which she powered her appliances, including a pump to get ground water from the aquifer. Where do you think her sewer goes? If you live in a shack in the middle of a mangrove, the cheapest solution is to dig a hole in the ground and let the raw sewer percolate through, in this case, the limestone contaminating the same groundwater from which you're drawing your water. Sadly, in the Yucatán peninsula, the Cenotes were also being polluted with black waters. Remember that people in the developing world don't want to deal with their shit just as much as you? I then offered to build her a compost toilet, but she refused, saying she was happy with her water flushing toilet and didn't need another one. In my opinion, the most ecological solution for the island's residents both rich and poor would be to have a dry, compost toilet saving water and turning feces into compost.

After the earthquake, some homeowners were given plastic tents which they set up in their front yards, adjacent to their destroyed homes. Some residences had simple latrine toilets apart from the homes which had survived the quake since they were light structures. These latrines are the worst kind of toilets. I'm offended when people mistake latrines for compost toilets. They are unsafe for the elderly and young because they are deep holes in the ground with unstable wood planks or poorly built slabs which sometimes break, and the user can fall through. Think of that scene from Slumdog Millionaire where the kid jumps into the shit, that's what a latrine looks like. Since there is no composting of the feces and they remain exposed, flies come and spread diseases.

The stench is also unbearable. There's one fact of living in places without toilets that, in my opinion, doesn't get discussed enough, if at all. It's demoralizing to live in a place that smells like shit. When you wake up and it smells horrible, you want to get the hell out of there. When you know that you're going back to a house or neighborhood and on the way, you have to cross a river of shit, it tends to make your mood shitty for a lack of a better word. I've noticed that when I'm at these places, I tend to downsize my plans, shorten my stays and build less toilets than I had hoped for. If you live in a place like this, slowly you can't help but give up on your dreams, aim lower and feel bad about yourself every day. This is the undignified life of billions of people today.

Unfortunately, we only get to see this from the perspective of those like me who are going there to help with toilets or whichever other humanitarian need. Sometimes, I feel awkward because it feels that I'm trying to promote myself. However, no matter how hard I try, the pictures on Instagram and posts on Facebook have become harder to post and write over the years. I don't even like giving qualms to beggars. When we do so, I think it's more to make us feel better about ourselves than really solving the root of poverty. But who has the time to sit down and talk to a beggar and help him or her? Who has the time to volunteer and help school street children?

What I love about the compost toilet is that it empowers the individual to take care of their own shit and consequently their own lives. It's the classic "teach a man how to fish" humanitarian approach. Since I've started using a dry toilet, I've become more aware of my own flaws, and that I need to take responsibility for my own actions. I've learned to apologize, speak the truth no matter how stinky it is and I'm more honest with myself. I believe that the compost toilet is a catalyst that will awaken us to our responsibilities to be conscious custodians of the planet earth and compassionate towards one another.

It was this understanding that gave me confidence to swim across the strait and it was also going to help me and Miguel build compost toilets for those who wanted one. First, we looked for a woodshop so that we could manufacture the wooden boxes which would house the plastic buckets and on which the toilet seats would rest. The hardware stores were filled with people buying material to fix their homes. There was always a line and many materials were out of stock. Luckily not many people bothered to buy toilet seats, so we ordered fifteen of their best seats available. We bought wood varnish, pesticide to keep termites out and delivered at the shop after agreeing that we would need fifteen boxes. Miguel took us to the edge of town where he knew a carpenter.

Coincidentally, the wood shop was near a migrant hostel built by a catholic priest to provide them with safe passage. All illegal migrants to the United States coming from Central and south America by foot must pass through Mexico and Oaxaca is the first state they reach. Once there, they board freight trains ominously known as *La Bestia* or The Beast, which is the same name as the official bullet proof car of the U.S. President.

These migrants are vulnerable to thieves, kidnappers and "coyotes" which are human traffickers. Women are sexually abused, and the migrants go through untold suffering on their journey to reach the United States in search of the American Dream. After having lived illegally in the U.S. and later obtaining dual citizenship, going to university and working in the *States*, I can say I've reached this same American dream they are chasing. I understand their struggle. However, after having left the US to build compost toilets in the countries these people are fleeing has put me in an odd situation. Should I warn them that they could one day be coming back to build toilets in México? Should I tell them to stop looking north and instead look down at the shit they're leaving behind on the train tracks? I don't think they would understand me because they are running away precisely from the shit in their own countries. I ponder all of this as I stare at the barbed wire cresting the walls protecting the migrant hostel.

It was Friday evening and Miguel's girlfriend was coming back. He arranged for me to stay with a professor friend of his, Mr. Ruben, in a neighboring city for the weekend while the wooden poop boxes or *cacajónes,* as César Añorve affectionately calls them, were being manufactured by the carpenters. I arrived at the professor's house at night and was welcomed to a very simple home by three elder men, one of them Sr. António, the professor's father. That night I slept outside, in a hammock tied to the tin roof columns. Apart from the mosquitoes, it was a nice breezy place despite the Isthmus being infamously hot all year round. They had a lot of jokes about how they lived in hell. In the middle of the night, around 2:30 am, my hammock was rocked violently, and I heard the roof snapping. I realized I had felt my first earthquake in life. I looked around at the other guys in their respective hammocks, but they didn't seem bothered, so I went back to sleep. They've grown used to earthquakes in this region.

Ruben or *El Maestro*, as he was called, was running for mayor of his small village and I witnessed the early days of his political campaign. He was happy to find more residents who needed compost toilets, perhaps he could get more votes that way. I was happy to count on his support. I had previously read that most of the aid money for sanitation gets diverted to water instead of toilets. Part of the reason is the negative connotation of toilets and politicians' tendency to shy away from projects that don't make them look good in the picture. *El Maestro* was glad to promote compost toilets and I welcomed his help. His campaign was picking up and we went to watch a rehearsal of the local traditional dance where men with masks whip and shout and the women wear pretty dresses. He met and shook hands with everyone as should every good candidate. Later we also took part in a religious parade along the city's cobbled streets and structurally condemned churches. These gatherings reminded me of my uncle and aunt's political campaigns in São Domingos, in the interior of Brazil. They were always being invited to barbecues, weddings, bars and the local river for to cool down from all the rallying. I also went to the river with Rúben and his supporters. I was also reminded of my change of heart regarding politics and the type of engineering which I want to. However, it was still inspiring to see *El Maestro* hopeful of a better future for his tiny village known for being hellish hot and prone to earthquakes.

On Monday, I returned to Ixtepec to get the wooden boxes and assemble the toilets by drilling holes and installing the seats. We bought buckets and sawdust to give each family. I also applied the varnish and pesticide because the carpenter ran out of time. Bad contract management on my part, I thought. Then, I held a workshop where I taught seven of the families in Ixtepec how to use the toilets and build a compost pile. Then, on Tuesday, I went back to Ruben's village to do the same, conduct a workshop with the remaining beneficiaries instructing them how to use the toilet and share the swim which enabled me to be there.

These cargo bikes in the picture above are common in Oaxaca. It's the main method of transportation for the simple men who are busy carrying materials, the groceries and even their family members in the front cargo bay. As a lover of bicycles, I appreciated this scene, having the dry toilet be carried away in a bicycle. I've had the idea of touring with a toilet and when I saw this in front of my eyes, I was inspired to one day cycle through some countryside with a compost toilet in tow. I couldn't resist the opportunity and asked for a ride.

In the end, I was content with my effort. We built and delivered fifteen toilets to families who had survived earthquakes. It's not much given the global population who needs toilets but if I learned anything in this swim, it's that the sanitation battle will be won one toilet at a time. I'm still working with toilets. I'm going back to México and I will build more and better toilets. I will work and continue dedicating my life to ecological sanitation until everyone hears and knows what a compost toilet is. As you've read in these pages, I'm persistent, and I'm happy with my work. I've found my life's calling and I'm giving it all that I have.

Have you gone to the toilet today? If not, then *Vai Cagar* and take care of your SHIT!

ACKNOWLEDGE-MENTS

Thank you Laura, I sincerely appreciate your patience and honesty especially at such a trying moment in your life. I appreciate the work your father did to enable swimmers like me to realize our dreams of swimming from Spain to Morocco. I applaud your for continuing his legacy. Muchísimas Gracias. I also want to thank Marianne Bruyeres for believing in and helping me at the beginning of the trip. I love you hermanita and I can't wait to travel with you again. Thank you, Francisco Pons, for letting me use the picture you took of Tita Lorens swimming in Playa Chica. I was behind her way out of frame. That place is where I trained most of the time. Also thank you, Polish photographer whose name I cannot find on fb anymore. I hope to find you and give you credit for the watermelon picture and other pictures you took of me at Playa Chica. Thank you Patty from "The Bike Revolution" bike shop in San Diego for sponsoring my ride. You're awesome man!

Thank you Clifbar for awarding me the Business with Purpose Scholarship in 2019. Who would have guessed that Chocolate Brownie Clifbar would have helped me not only to finish my swim but also win this scholarship? I appreciate your trust and I'm building a super portable compost toilet that is taking me on another great adventure.

Thank you Manu, this book would not have been possible without you.

I'm writing another book about the two toilets I built in India in early 2019 and I am also getting ready to start a twelve-month campaign of building and testing my toilets in the twelve countries with the worst sanitation indices in the world. This campaign will take me to obscure and less traveled places such as Burkina Faso Mali.

I hope that after reading this book, you can see why I have been working with toilets for the last three years. It hasn't been easy, but it's been worth it, and I want to keep going until everyone has access to sustainable sanitation. In order to continue my work, I've created a Patreon page where you can support me by making monthly donations and in exchange, I will share with you more details about my progress and challenges I encounter along the way.

Please follow me at www.patreon.com/toilets and on Instagram at @whereisaldoj

Until next time and take care of your shit!

Aldo Jansel

La Asociación de Cruce a Nado del Estrecho de Gibraltar

CERTIFICA

Que

ALDO DE PAULA JANSEL

Cruzó a nado el Estrecho de Gibraltar (neopreno) el 3 de **NOVIEMBRE 2017**
desde ISLA TARIFA (España) a PUNTA CIRES (Marruecos)
en 6h. 02m. Nadando una distancia de 8,2 millas (15,3 Km)
Acompañado por la embarcación COLUMBA

Presidenta de la Asociación

Laura Gutiérrez Díaz

[AJ1]